ENVIRONMENTAL RESEARCH ADVANCES

INVASIVE SPECIES MANAGEMENT

CONTROL OPTIONS, CONGRESSIONAL ISSUES AND MAJOR LAWS

Environmental Research Advances

Additional books in this series can be found on Nova's website under the Series tab.

Additional e-books in this series can be found on Nova's website under the eBooks tab.

ENVIRONMENTAL RESEARCH ADVANCES

INVASIVE SPECIES MANAGEMENT

CONTROL OPTIONS, CONGRESSIONAL ISSUES AND MAJOR LAWS

JEANNIE SAUNDERS
EDITOR

nova publishers
New York

Library of Congress Cataloging-in-Publication Data

ISBN: 978-1-63485-064-3

Published by Nova Science Publishers, Inc. † New York

CONTENTS

PREFACE

An "invasive" species (alternatively known as an alien, exotic, injurious, introduced or naturalized, non-native, nonindigenous, nuisance, or noxious species) refers to an animal or plant that is introduced into an environment where it is not native. The introduction of invasive species to the United States—whether deliberate or unintentional—from around the globe can pose a significant threat to native animal and plant communities, and may result in extinctions of native animals and plants, species disruptions as native and non-native species compete for limited resources, reduced biodiversity, and altered terrestrial or aquatic habitats. This can result in a range of economic, ecologic, and cultural losses, including reduced agricultural output from U.S. farms and ranches; degradation of U.S. waterways, coastal areas, national parks, and forests; and altered urban, suburban, and rural landscapes. Very broadly, the unanswered question regarding invasive species concerns whose responsibility it is to ensure economic integrity and ecological stability in response to the actual or potential impacts of invasive species, and at what cost. As this book shows, the current answer is not simple. It may depend on answers to many other questions: Is the introduction deliberate or accidental? Does it affect agriculture? By what pathway does the new species arrive? Is the potential harm from the species already known? Is the species already established in one area of the country? Finally, if the answers to any of these questions are unsatisfactory, what changes should be made? This book outlines the nature of the invasive species threat, the ability to predict invasions, methods of pest prevention or control, gaps in regulation, and options for congressional action.

In: Invasive Species Management
Editor: Jeannie Saunders
ISBN: 978-1-63485-064-3

Chapter 1

INVASIVE SPECIES: CONTROL OPTIONS AND ISSUES FOR CONGRESS*

M. Lynne Corn and Renée Johnson

SUMMARY

For the first few centuries after the arrival of Europeans in North America, plants and animals of many species were sent between the two land masses. The transfer of non-natives consisted not only of intentional westbound species ranging from pigs to dandelions but also of intentional eastbound species such as grey squirrels and tomatoes. And for those centuries, the remaining non-native species crossing the Atlantic, uninvited and often unwelcome, were ignored if they were noticed at all. They were joined by various species arriving deliberately or accidentally from Asia and Africa. The national focus on invasive species arose in the 19th century, primarily owing to losses in agriculture (due to weeds or plant diseases), the leading industry of the time. A few recently arrived invasive species, and estimates of adverse economic impacts exceeding $100 billion annually have sharpened that focus.

Very broadly, the unanswered question regarding invasive species concerns whose responsibility it is to ensure economic integrity and ecological stability in response to the actual or potential impacts of invasive species, and at what cost. As this report shows, the current

* This is an edited, reformatted and augmented version of a Congressional Research Service publication R44011, prepared for Members and Committees of Congress, dated April 30, 2015.

answer is not simple. It may depend on answers to many other questions: Is the introduction deliberate or accidental? Does it affect agriculture? By what pathway does the new species arrive? Is the potential harm from the species already known? Is the species already established in one area of the country? Finally, if the answers to any of these questions are unsatisfactory, what changes should be made?

The specific issue before Congress is whether new legislative authorities and funding are needed to address issues of invasive species and their increasing economic and ecological impacts on such disparate matters as power-plant operations, grazing lands, and coral fishes. Such legislation could affect domestic and international trade, tourism, industries dependent on importing nonnative species, those dependent on keeping them out, and finally, the variety of natural resources that have little direct economic value and yet affect the lives of a broad segment of the public.

In the century or so of congressional responses to invasive species, the usual approach has been an ad hoc attack on the particular problem, from impure seed stocks to Asian carp in the Chicago Sanitary and Ship Canal. A few notable attempts have begun to address specific pathways by which invasives arrive (e.g., ship ballast water), but no current law addresses the broad, general concern over non-native species and the variety of paths by which they enter this country. A 1998 executive order took a step in bringing together some of the current authorities and resources to address a problem that has expanded with both increasing world trade and travel and decreasing transit time for humans and cargo. Multiple bills have been introduced on this subject in recent Congresses.

There are two basic approaches to limiting the spread of invasive species: a species-by-species assessment of the risks or benefits of admitting or excluding species, and a policy based on controlling pathways of entry in which vigilance is maintained on incoming ballast tanks, cargo holds, packing materials, and similar vehicles for unwanted organisms. These two approaches may complement each other. Policymakers also may emphasize prevention over post hoc control or vice-versa, or they may adopt a combination of the two approaches.

INTRODUCTION

It is easy to find well-publicized examples of threats from invasive species: Burmese pythons multiply in South Florida, becoming top carnivores and killing large numbers of native species of reptiles, birds, and mammals. Formosan termites devastate living trees and historic buildings in New Orleans' French Quarter. Leafy spurge slashes the forage value of western grazing land, thereby creating precipitous drops in land value. To continue with a full list of the damaging effects of invasive plants and animals risks sounding like hyperbole.[1] So vast is this "bioinvasion" (as some have termed it) that only rough estimates can be made of the numbers of non-native species now in North America, much less the rest of the world.

However, it is unclear how best to manage or prevent these effects or what legislation would be needed to address these problems. Laws concerning wild plants and animals—whether native or introduced—do not form a comprehensive body at the federal level. The absence of a generalized federal responsibility in this area also circumscribes the federal role in addressing invasive species. Under the U.S. system, inherited from English legal tradition, the government regulates the taking of native wild animals generally and landowners control the native (and other) plants growing on their lands. Thus, colonial governments regulated native wild animals (to the extent there was regulation of the take of wild animals at all) and, after the U.S. Constitution was ratified, the states retained the rights they previously had as colonies to control the wildlife within their boundaries.[3]

With states retaining the right to regulate wildlife, the majority of wild plant and animal species are not federal responsibilities under current law. Moreover, because the problem of invasive species has continued to present itself as a series of seemingly disconnected crises, legislation has become a patchwork as each crisis has been addressed. This report outlines the nature of the invasive species threat, the ability to predict invasions, methods of pest prevention or control, gaps in regulation, and options for congressional action.

Invasive Species: What's in a Name?

Although this paper uses the term *invasive species,* many names have been used to describe species that are able to survive and reproduce outside the habitats where they evolved or where they have spread naturally. Among these terms are *alien, exotic, injurious, introduced, invasive, nonindigenous*, *non-native*, *nuisance*, and *noxious*. *Alien* tends to be applied to species from other regions that are also harmful and likely to proliferate wildly in their new habitat. *Injurious, invasive, nuisance*, and *noxious* generally are used synonymously and are not confined to species outside of their normal range. These terms are applied to species that proliferate wildly, whether native (like white-tailed deer and barnacles) or non-native (like Scotch broom and lionfish). In the legal arena, these terms are found in various laws and defined in various ways. Executive Order 13112 defines an invasive species as "an alien species whose introduction does or is likely to cause economic or environmental harm or harm to human health." This report primarily uses the term *invasive* to emphasize the rapid growth of such species.[2]

This paper focuses on those non-native species that have caused or seem likely to cause substantial economic or ecological harm. However, many species (e.g., cattle, olives, wheat, tulips, etc.) were introduced either with little known adverse ecological impact or with sufficient net benefits to make it likely that most people would not consider them harmful.

A Short History of Diffuse Legislation

Executive Order 13112

One executive order has been central in organizing the federal effort on invasive species. In response to rising concern, especially in southern and western states plus Hawaii, President Clinton issued Executive Order 13112 on invasive species (64 Federal Register 6183, February 8, 1999).4 The executive order sought to prevent the introduction of invasive species,5 provide for their control, and minimize their impacts through better coordination of federal agency efforts under a national invasive species management plan developed by an interagency National Invasive Species Council (NISC). The order directed all federal agencies to address invasive species concerns as well as refrain from actions likely to increase invasive species problems. NISC, supported by an advisory committee, was directed to develop recommendations for international cooperation, promote a network to document and monitor invasive species impacts, and encourage development of an information-sharing system on invasive species.

The range of relevant federal laws (as well as the federal agencies responsible for enforcing or administering them) is addressed in another Congressional Research Service (CRS) report,[6] but a short review of these laws is useful here. The laws addressing invasive species threats to agriculture—for centuries a well-developed North American industry whose risks from non-native invasions are relatively clear—tend to be more developed than laws protecting other industries or those protecting ecosystems. In consequence, agencies whose mission it is to address those risks also are better developed. Yet even the responsibilities to protect agriculture from invasives established in some U.S. regions but not others are diffuse, shared, or even lacking. Moreover, the enormous volume of trade makes the burden on federal inspection systems so severe as to permit only limited or cursory inspections and to force a strong reliance on self-reporting by importers of living and (apparently) nonliving cargo.

Laws protecting the natural ecosystems on which industries such as tourism, electric power, or city water supplies depend are far less developed. For example, a state agency that wishes to bring in a sport fish from another continent to benefit its anglers may face few obstacles in doing so, much less a burden of proof to show that the action will not harm other economic interests, natural resources, or ecosystems. Some laws, such as the Lacey Act,[7] force the burden of proof in the other direction: an import is deemed safe unless it is on a list of organisms known to be harmful.

A local or state government, faced with the recent arrival of a new invasive species—whether terrestrial, freshwater, or marine; plant or animal; agricultural pest or recreational nuisance—has no single source to query to begin its response or guide it through a maze of options. Moreover, federal help, especially any timely help in the weeks or months after initial discovery, is rare to nonexistent and focuses more on information and less on practical assistance. More than two decades ago, the Office of Technology Assessment wrote "The current Federal framework is a largely uncoordinated patchwork of laws, regulations, policies, and programs. Some focus on narrowly drawn problems. Many others peripherally address [nonindigenous species]. In general, present Federal efforts only partially match the problems at hand."[8]

This assessment largely holds true to this day, although since that time, the National Invasive Species Council (NISC; see "Executive Order 13112" text box, above) has addressed some aspects of the invasive species problem. It has taken steps toward sharing more information across governments and with the public. Based on its mandate under E.O. 13112, NISC asks specific agencies to take the lead in developing policies within their existing legislative mandates.[9] Through its 2008-2012 report, *Meeting the Invasive Species Challenge*, NISC outlined a set of actions to address the bulk of existing problems.[10] These actions include developing legislative proposals to fill gaps in current law. However, legislative efforts to date have tended to focus on well-established problems, such as the clear invasion of a single species, a specific pathway of introduction, or damage or risks to agriculture.

THREAT OF INVASIVE SPECIES

Various abundant invasive species have had severe economic impacts on U.S. industries and the natural environment. Damage varies by species and can span an enormous range of effects, including power outages; loss of farmland property value; contamination of grain; spread of disease; increases in operating costs; loss of irrigation water; collapse of buildings; competition with native plants; loss of sport, game, or endangered species; and ecosystem disturbance.

Some invasive plants have been notorious for years for causing both economic and ecological damage; kudzu, melaleuca, cordgrass, salt cedar, purple loosestrife, spotted knapweed, and Russian thistle are examples. Their damage includes lowering water tables, poisoning humans and livestock, decreasing crop yields, and increasing pest control costs. A serious infestation can cause substantial losses in property values.[11]

Invasive invertebrates also are well-known and include gypsy moths, Japanese beetles, Asian longhorn beetles, Asian tiger mosquitoes, fire ants, and Africanized honeybees. In addition, introduced vertebrate pests (e.g., walking catfish, lake trout, cane toads, monk parakeets, starlings, bighorn sheep, nutria, and rats) can have serious economic and ecological impacts. Introductions also include various human parasites and diseases.[12]

In some cases, the source of the introduction is either known or strongly suspected. Introducers of some species, such as kudzu, melaleuca, and starlings, actually intended that their imports proliferate in the wild. Their purposes ranged from the practical (kudzu for erosion control) to the quixotic (starlings for a purported desire to bring all of the birds mentioned in Shakespeare's works to the United States). Whatever the motive, society itself was the loser in the importers' Faustian bargains as the organisms proliferated to the detriment of the economy, native fauna and flora, and ecosystems.[13]

Introductions of non-native species may be intentional or unintentional. Some intentional introductions have produced benefits or at least low levels of harm, as in the case of the ring-necked pheasant (a game bird native to Asia) or honeybees (originally native to Eurasia). Unintentionally introduced species may be present for years or decades before their presence is even recognized and the harm that they do, if any, is measured or observed.[14] However, catastrophic results can come from both accidental introductions (e.g., sea lampreys in the Great Lakes) and intentional introductions (e.g., pigs in Hawaii).

Geographic Origins of Non-Native Species

Plants and animals tend to survive best in a new site that is similar to their original habitat. As result, the risk of a species becoming invasive could be affected by the ecological similarity between its origin and the U.S. region into which it is imported. Thus, the plants and animals of northern Europe, northern China, and New England are more likely to be a threat to the stability of each other's ecosystems than they are to be a threat to the fauna and flora of Miami or Singapore. Regions with similar climates and soils are tending toward increasing biological homogenization as species spread and new arrivals thrive at the expense of natives. This process reverses the normal evolutionary pattern of ever-greater species divergence between two geographically separated regions—from biodiversity to "biosimilarity"— as the new arrivals create a homogenized flora and fauna in the two regions.[15]

Near neighbors[16] are less likely than distant lands to be a source of pests. Canada, whose northern forests are contiguous with those of the United States, is not likely to be a source of forest pests because its forests have no meaningful biological barrier separating them from U.S. forests.

Consequently, if a native Canadian species could survive in a U.S. forest, it is probably already in one. Similarly, desert plants of the southwestern United States are probably shared with Mexico, and neither country's native desert fauna or flora poses much of a threat to the other.[18]

Exporting Invasive Species?

Although this report focuses on species that are not native the United States, this country is itself a source of species that threaten ecosystems elsewhere. On the "100 Worst" list of invasive species maintained by the International Union for the Conservation of Nature (IUCN), four are from the United States: eastern gray squirrel (*Sciurus carolinensis*), North American bullfrog (*Lithobates catesbeianus*), red-eared slider (*Trachemys scripta elegans*, a turtle), and largemouth bass (*Micropterus salmoides*). The grey squirrel of the eastern and northern states was introduced to Britain, where it has severely threatened populations of the English red squirrel. In 1958, Pan Am Airways and a local hotel introduced largemouth bass into scenic Lake Atitlan in Guatemala. This introduction, in the next 25 years, led to the crash of a sustainable crab fishery and of several native fish, contributed to the nearelimination of the commercial harvest of reeds used in local handicrafts, and forced the extinction of a bird found nowhere else. The harvest of the bass has since plummeted. The species has been introduced in other regions as well.

Even at the ends of the earth, human fecal coliform bacteria have contaminated scientific stations, including McMurdo Sound, the major U.S. scientific research base in Antarctica.[17] In an area thought of as earth's most nearly pristine environment, the bacterium *Clostridium perfringens* contaminates more than 80% of sea urchins near the undersea outfall of the untreated sewage from the base. (Treatment facilities are now in place.)

UNUSUALLY SUSCEPTIBLE HABITATS

Harmful non-native species occur throughout the United States, but some ecosystems are more susceptible to invasion than others. Mild climate, geographic isolation, disturbance of the natural landscape,[19] and a high rate of exposure to non-native species are all factors that can make a habitat particularly susceptible to invasion. Islands, remote lakes, and other long-isolated areas with unique plants and animals also are known to be particularly susceptible to invasive species.

Hawaii and Florida are perhaps the major examples of this phenomenon in the United States.

Both have many threatened and endangered species and, not coincidentally, a plethora of invasives. Both states were long isolated biologically and have large numbers of native species found nowhere else. The mild climates of Hawaii and Florida make it easier for the rich flora and fauna from other tropical and semitropical regions to survive and also make the states attractive to businesses that import, maintain, or breed non-native animals and plants, such as tropical fishes and ornamental plants. In Florida, one group compiled statistics on reported captures, kills, or encounters with exotic animals between January 2010 and October 2013.[20] The report included more than 56 Burmese pythons (see *Figure 1*), 15 pythons of other species, 1 gaboon viper, 304 monitor lizards, 100 bearded lizards, 1 emu, 1 kangaroo, 2 serval cats, 6 lemurs, and 5 capuchin monkeys. By focusing only on reported incidents rather than providing a systematic survey, the list likely understates the scope of invasive species in Florida.

Another factor putting some environments at risk is the sheer number of opportunities for new introductions.[21] Both Hawaii and Florida are major travel destinations and transportation hubs, so they are even more likely to be subjected to inadvertent introductions. Relative biological isolation, combined with easy transportation access also compounds the likelihood of invasion in aquatic habitats such as the Great Lakes.[22] Seaports, where many ships exchange ballast water, are at severe risk of invasions: even if only a tiny proportion of newly arriving non-native species survive in busy ports like San Francisco Bay or Chesapeake Bay, the actual number of successful invasive species may be large. The areas around airports, with increasing volumes of international traffic and tourism, are also at risk.

Burmese Python

The Burmese python (*Python molurus bivittatus*), as the name suggests, is native to Southeast Asia. It was brought to the United States for the pet trade and is thought to have been released or escaped from individual owners or pet stores. The first release or escape is not documented, but the first established population was reported in South Florida in 2000. Since then, the population has expanded throughout the Everglades and Central Florida, with a few reports farther north.

Source: Live Burmese python, Everglades National Park, Photo Files at http://www.nps.gov/akr/ photosmultimedia/photogallery.htm?id=3F4C9B3B-1DD8-B71C-078F792D3F1A85AF.

Notes: Larger snakes crossing roads may present a traffic hazard, particularly to small cars.

Figure 1. Burmese Python, Everglades National Park.

The extent to which winter temperatures limit the spread of the species is unclear. The snake is exceptionally large; the record length is 27 feet, and the record weight is 401 pounds. Evolving in an environment without such large snakes, native species of South Florida have few defenses, and a great variety of reptiles, birds, and mammals are prey.

Alligators are both prey and predator to the snakes, though alligator predation has been too low to control pythons. The species is now listed as injurious under the Lacey Act (see "Fish and Wildlife Service" below); its international and interstate shipment is severely restricted.

The habits of the Burmese python make eradication or even control extremely difficult. These snakes tend to stay in one place, waiting days or weeks for potential prey. After a large meal, a snake may remain hidden for weeks as it digests its meal. Although bounties have been offered for pythons, relatively few animals have been caught. There are no known examples of management action succeeding in controlling an established population. For more information, see http://www.invasivespeciesinfo.gov/animals/python.shtml; the website provides links to a wide array of federal, state, and other resources and information.

Pathways of Invasion

To some extent, pathways of invasion between countries can be predicted. For example, the propensity of brown tree snakes to hide in dark places has done much to focus attention on air stowaways. The arrival of a number of beetle species has played a similar role in focusing attention on pallet wood, packing crates, live plants, and airport warehouses as pathways and centers of biotic invasion. Live animals or plants may harbor microorganisms, parasites, or seeds that pose a danger to other species, even if the animal or plant itself does not survive in the wild.

In general, any arrival of living or untreated material could present a possible pathway for biotic invasion. A comprehensive review of possible pathways, their risks, options for control, and research needs is, to the authors' knowledge, currently lacking.[23]

Nonetheless, between or within countries, some pathways for species invasions are already well-known:

- Transportation corridors. Railroads, rivers, and highways can all spread invasive species. In the 19th century, the railroads over which cattle were transported were a major path for the establishment of new plants. In the 20th century, zebra mussels quickly escaped the drainage of the Great Lakes (probably via the Chicago Sanitary and Ship Canal) and began an invasion into the Mississippi River drainage, and Asian carp are repeating the process in the opposite direction. Interstate highways also help to spread invasive species such as weed seeds, freshwater bivalves (on trailered vehicles), and wood-boring beetles (in firewood).
- Intentional importation of non-native pets. The importer is hardly ever interested in seeing the imported organisms escape, preferring usually to sell, raise, or keep them instead. But once the specimens are sold, control is lost, and purchasers sometimes release unwanted animals into local lakes, streams, or forests, perhaps feeling they are being humane by releasing the animals.[24] The United States is a major importer of reptiles,[25] and though concern often has been for potential effects on host countries (e.g., loss of iguanas from Central America), there are issues regarding possible escapes, especially in southern states. A number of species are thought or known to have entered the United States as pets, or in association with pets: apple snails, goldfish, walking catfish, lionfish, budgerigars, ring-necked doves, and common pigeons, among others.[26]
- Landscaping plants. A number of woody invasive plants in the United States originally were introduced by the landscape industry.[27] A garden or greenhouse plant imported for horticulture may scatter shoots or seeds far more widely than expected. Water hyacinths, for example, were brought from South America in the late 19th century as pool ornaments. The plant now covers thousands of acres in the southern United States, plus parts of Cape Cod and California, as well as parts of Africa and Asia. As a result of this and similar escapes, the nursery industry has been subjected to increasing criticism, and some states are increasing their regulation of potentially invasive nursery plants.[28]
- Aquaculture. Many cultured aquatic species are not native. There is concern about the escape and establishment of cultivated species that may be harmful to native ones. Examples include the potential escape of Atlantic salmon in the Pacific Northwest; introduction of Asian oysters into Chesapeake Bay;[29] and the inadvertent introduction of

diseases, such as two diseases affecting oysters (called dermo and MSX), or of other pest species.

- Deliberate release for propagation in the wild. Deliberate release into the wild for propagation occurs for a variety of purposes. Several species of fish were released into the Colorado River for sport anglers, for example. (The continuing presence of these game fish has been one complicating factor in efforts to recover threatened and endangered species in and along the river, as well as to manage the river more naturally.) Salt cedar (or tamarisk) was introduced from Central Asia into the desert in the American Southwest in the early 19th century, in part to control erosion along river banks. The tree now forms dense thickets on more than 1 million acres of riparian habitat, and these thickets severely limit the growth of native cottonwood trees. To control the trees, in 2005 the Animal and Plant Health Inspection Service (APHIS, in the U.S. Department of Agriculture [USDA])[30] introduced the salt cedar leaf beetle (Diorhabda), which feeds exclusively on leaves of this tree, and the beetle began to reduce the populations of the tree. Unfortunately, it was determined that an endangered bird called the southwest willow flycatcher had adapted to the presence of this new tree and begun to nest in it, further complicating control of the tree as well as recovery of the endangered bird.[31]

Coordinating Science

Among recent efforts to coordinate scientific research on invasive species, a consortium of federal, state, and academic institutions formed the National Institute for Invasive Species Science (NIISS). Through its website, NIISS allows researchers to share a range of information on taxonomy, identification, geospatial data, and so on.[32] Its mission is

> to develop cooperative approaches for invasive species science that meet the urgent needs of land managers and the public. Administratively housed at the U.S. Geological Survey Fort Collins Science Center in Colorado, the National Institute of Invasive Species Science provides a hub for invasive species science collaboration, coordination, and integration across agencies and disciplines.... The mission of the National Institute of Invasive Species Science is to work with others to coordinate

data and research from many sources to predict and reduce the effects of harmful non-native plants, animals, and diseases in natural areas and throughout the United States.

However, the NIISS website is not intended for the general public and is difficult to use for such fundamental questions as the known distribution of invasives. A second organization, also a consortium, is the North American Invasive Species Network (NAISN). It is a

> non-profit organization that was formed in 2010 by university and government scientists from across North America.... The countries of Mexico and Canada participate as NAISN members through a Memorandum of Understanding.... [Members are] regional university centers and institutes, government institutions, non-profit organizations, research labs, and/or other groups and individuals with invasive species interests and qualifications that are valuable to the mission of NAISN.... NAISN aims to unify and connect existing regional invasive species efforts into a single network ... to help current invasive species management and prevention efforts across the continent.

Its website (http://www.naisn.org/) is designed for a wider audience than NIISS. It offers videos demonstrating the characteristics of several dozen invasive plants and lists services offered by the consortium; the services focus on data collection and coordination among agencies.[33]

PREDICTING AN INVASION: BLACK AND WHITE?

Federal laws have tended to focus on *black lists* (anything not on the list is allowed) in contrast to a *white list* (anything not on the list is excluded). (See summary of various approaches in *Table 1*.) Each requires some, or even considerable, knowledge of the species to be listed in order to predict the likelihood of invasion. The types of lists are discussed below.

Methods for Predicting Invasiveness

A central dilemma in making any list is the difficulty in predicting a species' ability to invade. What characteristics of seed dispersal, nesting, food and host preferences, etc., are most likely to lead to exuberant proliferation and result in economic and ecological harm? For example, it seems logical that, all else being equal, plants producing many seeds will be more invasive than those producing few seeds. The problem is that "all else being equal" rarely *is* equal—a variety of factors affect invasiveness. The abundant seed producer may require a special pollinating insect; the newly arrived plant with few seeds may leave behind its major herbivores, and so on. A host of other factors may complicate prediction.

Nonetheless, there are a few characteristics that may serve as predictors. One example is *propagule pressure*.[34] This term describes those species that attempt invasions (somehow arrive in large numbers) most frequently and/or with the largest number of arriving individuals and therefore are more likely to be successful invaders. A second strategy to model invasiveness is Hazard Analysis and Critical Control Points (HACCP), a method to analyze the critical points at which an invasion may be prevented or stopped.[35] Scientists continue to model other factors that might increase invasiveness. However, no evidence to date has identified a suite of features that seems to be a reliable predictor of invasiveness, and thus many experts view all importations as suspect.

Black Lists

A black list can be prepared in various ways, but it usually is made up of species already shown to cause serious damage to fisheries, endangered species, or (especially) agriculture. This evidence may be based on experience with the species domestically or in other countries. Preventing the spread of the species after it enters the United States may have to rely on public education, penalties for shippers, monitoring, and other means. In general, black lists require time to gather information on the damage created by the species and then proceed through rule-making. This approach allows more flexibility for industries that depend on the importation of new species of plants or animals. Black lists do not readily address introductions by persons who are unaware that they are bringing in non-native organisms.

White Lists

With the white list approach, there is an attempt to predict potential harm before a species' arrival. The prediction would be based on known characteristics of a species, such as how it reproduces, the number of seeds or offspring, etc. Species that are not predicted to cause harm would be added to a white list of allowable species, and any species not on the list would be excluded. The mongoose, for example, has a history of becoming a pest on islands where it has been introduced. It seems unlikely that the mongoose would ever be placed on a (white) list of allowable species. In contrast, new varieties of orchids, sheep, tulips, or any other species with a long track record in this country would likely gain admission.

Table 1. Intentional and Unintentional Introductions: Potential for Control

Nature of Introduction	Potential for Control	Potential for Control	Potential for Control
(examples)	with White List	with Black List	by Pathway
Intentional Importation and Release (St. Johnswort, mongoose)	Yes	Yes	Yes
Intentional Importation, Unplanned Release	Yes	Yes, if coupled with post-purchase control	Yes, if coupled with post-purchase control
(Burmese python, giant African snail)		(education, penalty, monitoring, etc.)	(education, penalty, monitoring, etc.)
Unintended Importation (emerald ash borer, brown marmorated stinkbug)	Possible	Unlikely	Yes, if pathway regulation is adequate
Unintended Importation via High-Risk Pathways (ballast water with zebra mussel, pallet wood with	Possible	Unlikely	Yes, if pathway regulation and compliance is adequate
Asian longhorn beetle)			
Species Entering via Pathways Still Unknown	No	No	No, until pathway is discovered

Source: Prepared by the Congressional Research Service.

No List

The existence of a list, whether white or black, implies that people actually know they are importing living organisms. An effort to prevent unintentional introductions would be compatible with any shade of list or no list at all. Potential pathways continue to be discovered, and there are likely other pathways yet to be determined.

BENEFITS FROM NON-NATIVE SPECIES

Although the ecological damage from some non-native species can be great, many such species are beneficial. (For many species, the economic impacts are simply unknown.) Some industries rely heavily on non-natives. For example, very few food plants and animals in the United States are native to the areas where they are now grown. [36] In addition to agriculture, the nursery, aquaculture, and pet industries rely significantly on non-native species.

A wide variety of intentional introductions have had effects that, even if harmful to natural ecosystems or biodiversity, have produced economic benefits sufficient to cause acceptance of collateral damage. These include such food sources as cattle, chickens, honeybees, wheat, kiwi fruit, and soybeans and such ornamentals as tulips, chrysanthemums, and dawn redwoods, to name a few. In each case, the introduction of these species was intentional and their propagation was more or less controlled. The economic benefits conveyed by these species are vast and probably exceed many estimates for the annual costs of invasive species.[37] U.S. agriculture would have a far different appearance if it were limited to the several dozen food crops known to have been cultivated in North America before 1492 rather than the hundreds of crops grown today. These non-native crops and their benefits are not the focus of this report but should not be forgotten in discussions of those imported species that cause serious harm.

Invasive Species Prevention: Stopping at the Border

The first—and perhaps only—line of defense against invasive species is preventing introductions into the country because controlling invasive species, once established, has been difficult. Prevention usually is economically advantageous because established invasive species can rarely be eradicated. Furthermore, controlling these species to acceptable levels, when or if possible, may be an expensive and endless task.

Guidelines for Release into New Regions or Nations

Some international organizations have developed guidelines and codes of practice for the deliberate release of non-native organisms. The World Organisation for Animal Health,[38] the Antarctic Treaty Secretariat,[39] and the International Council for the Exploration of the Sea,[40] for example, have prepared such guidelines to assist regulatory bodies and other groups to determine whether an introduction is justified and then to advise them on what to do after an introduction is approved. Such guidelines focus on hazard identification, risk assessment, risk management, and risk communication. They provide a conceptual framework for determining whether the risk of introduction is acceptable, and they suggest any necessary quarantines, monitoring, and/or adaptive management if an introduction is approved.

Quarantines and Inspections

Quarantines and inspections are methods of prevention by which the entry of non-native species through specific pathways might be controlled. The quarantine approach requires, first, that species be recognized as pests or that pathways be recognized as likely to transfer pests. Second, there must be a prohibition on entry, either of particular organisms or of particular types of cargo.[41] Quarantines operate basically on either of two strategies:

- investing in strict control at points of entry to prevent organisms from leaving the quarantine area (with the risk that it could be too late to prevent escape) or

- attempting to control before the organisms arrive in the country (i.e., trying to act on the source or point of export or regulating the pathway of import).

Cargo and luggage inspections at ports of entry are examples of the former. A quarantine also may be posted in an area where an invasive species has been introduced to prevent its further spread and promote its eradication. An example of the latter occurs when contaminated ballast water is treated or removed at sea and before entry into port to eliminate potential invaders.

Quarantines and inspections for agricultural pests are a major responsibility of APHIS. In addition, various states (particularly California, Arizona, and Florida) and territories conduct their own inspection programs aimed at agricultural pests. However, inspections and quarantines may be inappropriate for many species, especially species that enter through uncontrolled or unrecognized pathways and species that are not yet recognized as pests.

BASIC METHODS OF INVASIVE SPECIES CONTROL

If inspections and quarantines fail to keep an invasive plant or animal out of a region and a species becomes established, the problem shifts to control of the species, which includes preventing its spread between local areas and beyond any established perimeter. Control of invasive species is divided into two related tasks: eradication, where possible, and reduction to manageable/tolerable levels[42] where eradication is not possible. Key to this effort is early detection and rapid response (EDRR).[43] With EDRR, eradicating very small populations before they become established may be possible and is more likely if many methods, such as those described below, are used intensively and in combination, including treating outlying populations as soon as they are discovered. Few control methods, if any, promise eradication if a species is well-established, but several methods, used in combination and continuously, might reduce a target species' population to tolerable levels.[44]

Several methods may be used to address an invasive species population, including cultural controls, mechanical controls, baits and attractants, biological controls, chemical controls, and bounties. To apply any of these strategies, substantial knowledge of the target species' behavior, biochemistry, dietary preferences, diseases, or other aspects of its biology is usually essential.[45] Combining pest control with an understanding of the underlying ecology of the species is central to integrated pest management (IPM), which may be defined as:[46]

> an ecosystem-based strategy that focuses on long-term prevention of pests or their damage through a combination of techniques such as biological control, habitat manipulation, modification of cultural practices, and use of resistant varieties. Pesticides are used only after monitoring indicates they are needed according to established guidelines, and treatments are made with the goal of removing only the target organism. Pest control materials are selected and applied in a manner that minimizes risks to human health, beneficial and nontarget organisms, and the environment.

Generally "the most effective, long-term way to manage pests is by using a combination of methods that work better together than separately."[47] The strategies described below overlap somewhat, both in concept and in practice, but their differences in timing or breadth of application make distinctions useful.

Preventing Dispersal after Entry to the United States

After introduction, prevention means taking steps early to limit the spread of an invasive species that has begun to escape confinement (from a few lakes, one airport, one farm, one region, etc.). Prevention practices vary depending on the species and surrounding environment, and they may include more than one of the methods described below. Often this may simply require "removing the pests' sources of food, water, and shelter, or blocking their access into buildings or plants" to prevent population growth.[48] This also may involve implementing certain best management practices for invasive species.[49]

For example, to prevent the spread of invasive species in forested areas, the U.S. Forest Service encourages forest land managers generally to identify suspect plants or animals, prevent the introduction of seeds/eggs/organisms into an area; detect and promptly eradicate new outbreaks; minimize disturbance of desirable vegetation; build up native species' populations; periodically inspect high-risk areas (such as access points, transportation corridors, and disturbed ground); manage stand density and growing conditions in forested stands; revegetate disturbed sites; and periodically evaluate management protocols.[50] Prevention of invasive aquatic species that are just becoming established in a limited area may involve equally wide-ranging efforts.[51]

Controls Designed for Confined Spaces

Lethal substances can be used to target pests in confined areas or to prevent them from crossing a geographic bottleneck. Methods for use in confined spaces are often—but not exclusively— pesticides. (See also "Chemical Controls.") These methods can be used either to create pest-free "islands" in a zone of infestation or to prevent invasive species from leaving an infested area via boxes, cargo holds,[52] etc. Some species are known to avoid certain substances, such as tear gas or gasoline. Obviously, some of these substances can be used only in areas where human access is infrequent. In a confined body of water, a compound called rotenone can be spread to kill invasive fish—possibly involving a short-term sacrifice of resident fish. Light is known to repel some nocturnal animals. Submerged surfaces have been electrified at water and power facilities to discourage the settling of zebra mussel larvae and on ship hulls to inhibit barnacle settlement. Barriers can be used to discourage access to new habitats, such as the electrical barrier in the Chicago Sanitary and Ship Canal intended to prevent or reduce the dispersal of invasive Asian carp.[53]

Cultural Controls

Cultural controls refer to agricultural production practices that modify a pest's environment or habitat to reduce the pest's spread by reducing its ability to survive, disperse, establish, or reproduce. Examples include crop rotation techniques;[54] intercropping (also companion or mixed cropping); managed application of water or fertilizer; improved sanitation and hygiene; timed plantings and harvests; purchase of certified plants (grown under sterile or quarantined conditions); soil solarization;[55] tilling or mulching weeds; use of resistant plant and seed varieties (such as resistance to diseases, nematodes, or certain weeds); practices that change soil pH or fertility levels; irrigation practices; and use of beneficial predators (such as chickens or ladybugs).[56] Some cultural control methods, although minimal in cost and equipment, may require an additional commitment of labor and are unlikely to be effective unless combined with other control methods.

Mechanical Controls

Mechanical and physical controls refer to a variety of techniques and include manual controls (such as hand-pulling weeds or physically removing a plant or animal); mowing; use of mulches for weed management; barriers such as screens; the use of heavy machinery such as harvesters and shredders designed to cut, shred, crush, lift, transport, and remove pests (such as stands of aquatic plants and associated organic material); use of a specific tool for removal of specific pests (including traps); and other types of physical methods of pest control, such as hot water or steam treatment and steam sterilization of the soil for disease management. Such actions are intended either to kill a pest directly or to make the surrounding environment unsuitable. Mechanical controls also are used to prevent the further spreading of established invasive species, and they include cleaning the equipment (e.g., during highway construction) and using certified weed-free seed and feed (e.g., weed seeds have been mechanically excluded or removed).

White-Nose Syndrome

White-nose syndrome (WNS) is a newly discovered disease of bats. It is caused by a fungus called *Pseudogymnoascus destructans*. The fungus appears as a white, powder-like substance on an affected bat's nose, wings, tail, or ears. WNS was first discovered in bat caves in the winter of 2006-2007 in upstate New York, and it has since spread rapidly to bats in 29 states and 5 Canadian provinces, evincing extremely high mortality rates, sometimes approaching 100%. The disease was unknown in North America before 2006 but is frequently found in Europe, where bats appear to be more resistant to the disease. Infection is thought to be spread via contact between bats and by contaminated caving equipment. Affected bats appear to be starving, and leave their caves while weather is still too cold to permit survival so that in the springtime, cave entrances may be littered with dead bats. Multiple species are susceptible, including the endangered little brown bat (*Myotis lucifugus*). While WNS is not the only threat, it plays the major role in the proposed listing of the northern long-eared Bat (*Myotis septentrionalis*) as endangered under the Endangered Species Act. There is no known cure, although education efforts are being used to slow the spread of the disease.

Economic and health effects could be profound. For example, bats have a critical role in controlling insects, including moths that affect crops;

reduced bat populations could increase crop loss and the potential cost to farmers of more insecticides. Bats also play some role in preying on mosquitos, including some species known as disease vectors for humans or other species. Universities and federal agencies are carrying out multiple lines of research in the search for a solution.

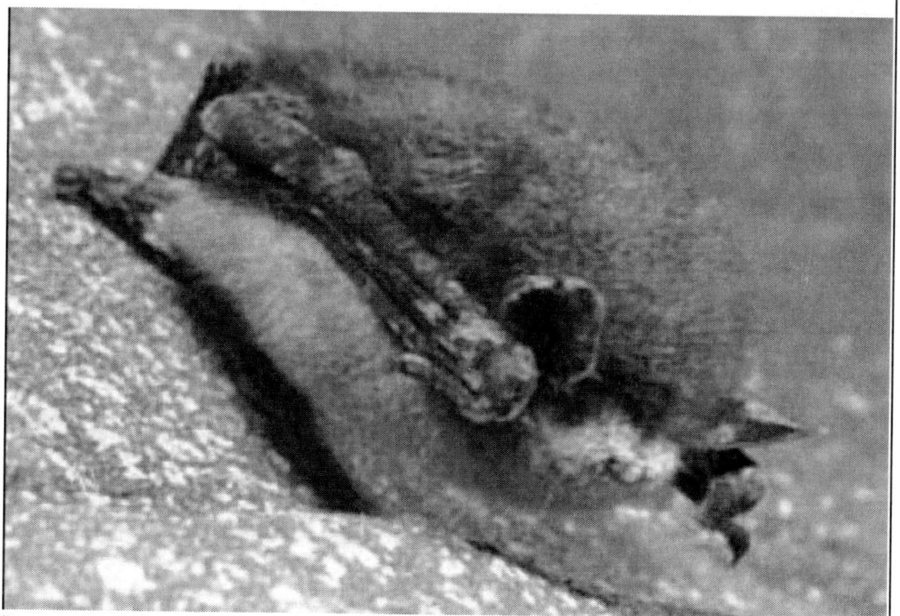

Source: http://www.invasivespeciesinfo.gov/microbes/wns_child.shtml; photo by Al Hicks, New York Department of Environmental Conservation.

Notes: White muzzle of this endangered bat is typical of the usually fatal fungal disease; this animal also shows white on its wings and ears.

Figure 2.White-Nose Syndrome in Little Brown Bat.

Like cultural control methods, some of these methods may require expensive specialized equipment or a substantial commitment of labor to be effective. For example, mechanical harvesters may be used in the management of invasive aquatic vegetation, such as hydrilla and water hyacinth, but are ineffective for control of these species on large bodies of water.

The use of traps for rodent control also may be limited by cost, as well as by the time required to service traps and the inability of traps to control target species over large areas. Various trap designs are available, and most are used in combination with some type of bait or attractant, such as prey or attractive scents (see "Baits and Attractants"). For confined areas such as cargo holds and buildings, traps may be successful. However, traps have drawbacks in open situations with either abundant alternative foods or very low target species densities.[57] In addition, traps may affect nontarget organisms. Traps are relatively safe to use, although they require some care when trapped individuals are killed and removed. Sticky traps have been used on rodents and cockroaches. Invasive Chinese mitten crabs have been trapped at irrigation screens during their downstream migration to spawn in saltwater.

Baits and Attractants

Baits and attractants are used to draw individuals of a target species toward a potential source of food or mates (see also discussion of pheromones under "Chemical Controls"), where the target species can be counted, trapped, killed, or studied. Difficulties with baits and attractants commonly include sustaining a long-term monitoring effort and preventing harm to nontarget species. Baits and attractants seem most promising when the area needing protection is well-defined with clear boundaries and has a significant density of the target species.

Biological Controls

A biological control organism competes with, preys on, parasitizes, or causes disease in a targeted pest species. Ideally, biological control agents attack the target species and no others.[58] Considerable knowledge of both the target species' and the control organism's basic biology and ecology is necessary to select a suitable control. Together, individual state laws and APHIS[59] regulate the introduction of biological control organisms, and USDA's Agricultural Research Service administers a Biological Control Documentation Program.

A particular concern with biological control organisms is that they might eat or infect nontarget species once target species are sparse or eradicated. A historical example of this problem is mongoose introductions. In the 1600s, mongooses were introduced in Puerto Rico to eradicate rats, which they did with great success. Unfortunately, mongooses proliferated and began to eat a variety of birds and other native animals. In light of such problems, vertebrate animals with broad feeding habits are seldom, if ever, used today as biological control agents.

Fieldwork in a species' native habitat usually is necessary to identify an effective disease or parasite. Although the requisite research might be expensive or slow, biological agents may offer long-term control, if not eradication. The use of alligator weed flea beetles (*Agasicles hygrophila*) for control of alligator weed (*Alternanthera philoxeroides*) is one example of this tactic.

Natural biological control also can occur through adaptive ecosystem response by native species to invasive species. An example in Oregon is larvae of the native defoliating butterfly, *Vanessa cardui*; they feed not only on their normal broad range of host plants but also on introduced thistles, *Cirsium arvense* and *C. vulgare*.

Other forms for biological control may involve planting competing vegetation and managing livestock grazing. For example, grazing by sheep and goats can be an effective management tool for controlling leafy spurge.

Chemical Controls[60]

In integrated pest management (IPM) strategies, pesticides may be considered as part of the strategy but generally are a last resort, after other nonchemical methods have been exhausted or proved ineffective or unavailable. Pesticides are used "only when needed and in combination with other approaches for more effective, long-term control" and "are selected and applied in a way that minimizes their possible harm" to humans and other organisms, as well as the environment.[61]

Chemical control agents can be subdivided into those derived from manufactured (conventional) or natural (biological) sources.

- *Conventional.* Where chemical control is an option, pesticides affecting or controlling only one or a group of related species are strongly preferable because broadly toxic substances risk substantial harm to nontarget species. For example, TFM (3-trifluoromethyl-4-nitrophenol) is very specific in its toxicity to larval lampreys. Similarly, a variety of aquatic herbicides can be used specifically for the control of hydrilla and water hyacinth. In contrast, methyl bromide is broadly fatal to many species.[62] However, even if pesticides are highly specific, safety precautions often suggest the use of the chemical agent in conjunction with baits to reduce risks to children, pets, and other nontarget organisms.

- *Biological.* Biopesticides are derived from natural materials, such as animals, plants, bacteria, and certain minerals. In 2014, EPA reports there were 430 registered biopesticide active ingredients and 1,230 actively registered biopesticide products.[63] Biopesticides can be divided into three major classes:
    - biochemical pesticides are naturally occurring substances (e.g., pheromones) that control pests by nontoxic mechanisms (e.g., interfering with finding mates);[64]
    - microbial pesticides contain a microorganism (e.g., a bacterium, fungus, virus, or protozoan) as the active ingredient, such as various types of the bacterium Bacillus thuringiensis (Bt) used to control certain insects harmful to cabbages, potatoes, and other crops;[65] and
    - plant-incorporated protectants are pesticides that plants produce from genetic material that has been added to the plant, as when the gene for the Bt pesticidal protein is introduced into a plant's own genetic material, causing the plant to manufacture the substance that destroys pests.[66]

Although biopesticides tend to pose fewer risks than conventional pesticides (they tend to be less toxic, usually are effective in very small quantities, often decompose quickly, and generally affect only the target pest and related organisms), users need to know a great deal about pests to employ them effectively. Their use is commonly part of IPM.

Bounties and Commercial Exploitation

Under a bounty system, someone is paid to catch and kill the target species. High bounties may be necessary for a substantial effect on the target species' population. For example, a 2013 bounty hunt for Burmese pythons in Florida produced only about 50 captures, although more than 1,500 persons sought licenses, according to press reports.[67] Moreover, paying bounties can create a market incentive to produce or introduce more of the invasive species—a particular risk when a population dwindles to very low levels and prices go up or bounties are increased. In August 1999, the California

Department of Fish and Game decided against permitting the commercial harvest of invasive Chinese mitten crabs, concluding that such harvest would not contribute to controlling this species and might encourage further introductions.

Site Removal

One unusually drastic method is to remove or poison each site where the invasive species lives. This option works best when the sites are isolated or at least well-defined. For example, the Asian longhorn beetle currently is being controlled by removing all trees on which the pests might feed in neighborhoods where they have been found. The species has been eradicated in New Jersey and Illinois using these methods in combination with other measures. Other states also are employing the same tactic.[68] In addition, site removal has been used in California to eradicate hydrilla by draining small ponds and filling their depressions with earth. Such a strategy becomes unworkable if a pest is widely dispersed.

Use of Other Species to Detect Invasives

Rather than using one species to control another, a slightly different technique is to use one species to detect the presence of an invasive species, after which other methods are used to kill individual organisms. Commonly, this approach involves training dogs to detect the scent of the target species. Dogs have been used to detect Burmese pythons in Florida[69] and spotted knapweed,[70] among other species. Training of such dogs is expensive, and cost may limit their use. However, their utility at high-risk entry points (airports, cargo terminals, dockyards) decreases their cost per find and may prevent the entry of invasive species at the most efficient phase: before the invasive species has entered.

Issues for Congress: Actions and Approaches

Current federal laws concerning invasive species form a patchwork, stronger in certain areas, such as agriculture and ballast water, and weaker or absent in other areas.[71] Current laws do not clearly address

- prevention of biological invasion across foreseeable pathways (with the exception of ship ballast water) or
- early detection and rapid response before the establishment of the new species. At the latter event, the focus of effort shifts from less expensive prevention to more expensive and less efficient control.[72]

Coordination of current efforts alone means that, where there are gaps due to lack of coverage by existing laws or agency jurisdiction, those gaps will remain. Congress could choose to address these gaps either by explicitly delegating such authority to the President or by crafting legislation.

Federal Agency Actions: Patchwork and Gaps

The National Invasive Species Council (NISC) has become the focus for federal efforts to control and prevent invasive species affecting a broad range of industries or ecosystems. However, many of the past shortages of authorities or personnel that have hampered efforts to limit the entrance of and damage from invasive species remain. Without clear federal legal authority to protect the nation's ecosystems (except in agriculture), the information sharing under NISC and the National Institute for Invasive Species Science (NIISS) takes on even greater importance. Simply informing responsible parties of the risks of their proposed actions may be one of the few vehicles available to deter potentially damaging releases due to gaps in the coverage of these agencies or their statutes.[73] Although many agencies have some role in addressing invasive species, four agencies—APHIS, the Army Corps of Engineers, the Fish and Wildlife Service, and the National Marine Fisheries Service—have had major roles for many years. Yet even for these four agencies, important gaps remain; the gaps in their authorities are described below.

Animal and Plant Health Inspection Service

Although APHIS oversees the importation of plants or animals, its authority to regulate living plants or animals once they are admitted is more limited. Specifically, it does not regulate the release of a species into the wild once they have been admitted, unless that species is already designated a noxious weed or disease carrier. For example, APHIS could inspect imported emus or ostriches for the sake of protecting domestic poultry from foreign diseases. However, after the markets for these birds crashed some years ago and many were released by emu ranchers, APHIS did not have authority to control the release of this non-native species. Nor does APHIS regulate the sale of species that are already well-established. For example, although English ivy is widely recognized as an invasive plant and harmful to forests, its sale does not violate federal law.[74] The sheer volume of plant and animal imports into the United States, as well as ever-increasing trade, raises practical questions about APHIS's ability to maintain its oversight role.

Army Corps of Engineers

The Corps attempts to control noxious aquatic plant species (e.g., hydrilla) and other impediments to inland navigation.[75] Yet the Corps has no authority to prevent or regulate the release of such pests in the first place. The hydrilla infestations in Florida and the Potomac River, for example, very likely were caused by aquarium hobbyists dumping fish tank contents, but it is unclear whether the dumpers violated any state or federal laws. Similarly, the Corps did not regulate the import of Asian carp by fish aquaculturists, but it now has a major responsibility to manage certain Corps locks in Illinois to prevent the spread of this species from the Mississippi River basin into the Great Lakes.[76]

Fish and Wildlife Service

Fish and Wildlife Service (FWS) does not have authority to provide general protection for the country's ecosystems from invasive species, nor even to protect most species popular with hunters and anglers. Only if the species were already regulated under the Lacey Act,[77] or if FWS knew that the introduced species might harm a species protected under the Endangered Species Act,[78] would it have any authority to stop the importation and release. Instead, FWS waits for evidence of injury from the species to place it under regulation through provisions of the Lacey Act. It does not currently have emergency authority under this law. To make listings under the Lacey Act more expeditious, FWS proposed a categorical exclusion under the National Environmental Policy Act (NEPA).[79] The proposal would have listings continue to be subject to the other "legal requirements of the Lacey Act, the Regulatory Flexibility Act, and other required determinations for all injurious [species] rulemakings" as well as other requirements of NEPA.[80] The regulation has not been finalized.

National Marine Fisheries Service

On inland waters, the National Marine Fisheries Service (NMFS) has little regulatory authority because its responsibilities largely are confined to demonstrable effects on anadromous species protected under the Endangered Species Act. Moreover, its authority to prevent harmful introductions into marine ecosystems appears to be nonexistent. The dumping of invasive lionfish into the waters of the Atlantic Ocean and Gulf of Mexico, for example, appears to have violated no federal law.

Approaches to Regulation: Species-by-Species vs. Pathways

Legislation to address invasive species could take a species-by-species approach, a pathway approach, or a combination of the two. The species approach implicitly assumes knowledge about a species' risk (black list) or safety (white list). A central dilemma, however, is the difficulty in making this prediction.[81] Moreover, this approach assumes knowledge that a particular species actually is being imported. A pathway approach does not assume knowledge about any particular species, only that a particular set of circumstances favors the arrival of unwanted organisms.

Some agencies, including APHIS and FWS, currently analyze the risk presented by particular species that may become invasive. However, addressing the multitude of agricultural pests relies on scarce agency resources at APHIS. In addition, some have criticized FWS for its slow response to the blacklisting of species under the Lacey Act. Both agencies arguably would improve their effectiveness if they could provide faster assessments of either species or pathways so that they could direct resources to the most critical areas.

Regulation by pathway is an approach suited to unintentional or unknowing introductions, as no list needs to be created. Among the most comprehensive pathway approaches to date is the Non-indigenous Aquatic Nuisance Prevention and Control Act.[82] Its goals put prevention on an equal or higher footing compared with control of species that are already established. It requires the participation of several federal agencies, promotes research, and implements regulations on the mid-ocean exchange of ballast water and other measures to exclude invasives from U.S. ports.

Legislative or Policy Options

A review of writings by various specialists in this field suggests additional areas that might be explored by policymakers.[83] Over the years, Congress has considered many issues related to particular invasive species. The list of options below is compiled from many sources and focuses only on topics that cover a broad range of species, pathways, or agencies.

- *Research to identify pathways.* With the exception of agricultural threats, few comprehensive reviews exist to identify pathways providing the greatest risk of invasives. Research goals in this area might overlap with research designed to prevent certain kinds of security threats, and might benefit from cooperation with agencies involved in antiterrorism programs.
- *Expert review of planned releases.* Panels of experts might be created to analyze risks and make recommendations on planned releases by governmental or nongovernmental sources into any environment in which the species are not native. According to NISC,[84] such steps are planned and, in the case of plants that may affect agriculture, are in progress at APHIS. Although such a panel could not have sounded the alarm on the unauthorized release of hydrilla into the Potomac, for instance, it could provide a public warning on planned releases of exotic grasses by federal agencies or of non-native game fish. Exotic plants are proposed as feedstocks for biofuels and provide a current example of a release that might benefit from such reviews.[85] The use of the expertise of federal and other scientists and managers, if it prevented even a few ill-advised introductions, might be a cost-effective option.
- *Educational campaign.* An educational campaign, possibly aimed in part at children, to prevent simple, inadvertent acts by the public might play a role in preventing some types of invasive species introductions.[86] Preventing releases of exotic pets and aquarium species after the point of sale might be particularly susceptible to this type of approach.
- *Warning list.* An informational *warning list* (or *gray list)* of species might be created by the collaboration of federal and state agencies. The warning list might include species currently restricted under state laws, species thought to be newly arrived from other countries, and other species felt to merit special attention by regulators.[87] Although a gray list would lack regulatory force, at least without legislation at the federal level, it could be designed to provide information on species whose eradication or control is in its early phases. Unification of data

and reporting from many agencies would add greatly to the utility of such a list.

- *Review of industries dependent on importing and transferring non-native species.* Such a review could include a focus on cooperative methods to reduce introductions or releases after the point of sale. The focus of past efforts has tended to be on the entry of these species into the United States. To protect their businesses, import-dependent industries naturally have tried to reduce current obstacles and prevent imposition of new ones. In this effort, the pet, hobby aquarium, and nursery businesses have been relatively successful. Yet there are other avenues to reduce risk besides prohibition. These avenues might include incentives for the sale of sterile animals or plants only[88] or efforts to create point of-sale educational programs about the risk of releasing pets or plants into the wild and any penalties for doing so.[89]
- *Multi-agency federal or cooperative center for "first strike" prevention and control.* Congress could authorize the creation of a first strike center along the lines of the National Interagency Fire Center.[90] Since the creation of NISC, agencies have begun to respond across a broad front in the days, weeks, or months after an invasion is discovered. The prompt notification and agency attention to the discovery of northern snakeheads in Maryland is an example of such a response. (Ultimately, the effort was too late, and the fish spread widely in the Chesapeake Bay area.) However, although constraints on coordination and jurisdiction interfere with prompt responses less than in the past, much more progress is possible. NISC is beginning to model its efforts on interagency fire management, a federal program that has long faced similar issues. It seems possible that a similar center devoted to first strike prevention and control of invasive species, regardless of affected industry, ecosystem, or lead agency, could provide critical support at a time when eradication of a new animal or plant invader might still be possible. In recent years, the National Park Service (NPS) has adopted roughly this approach for plant pests. Emergency Plant Management Teams address not only crises but also ongoing problems.[91] While the current 16 teams work primarily on NPS lands, they also may work with adjoining landowners. Expansion of this program, perhaps with multi-agency teams, could provide a more efficient method of addressing new or incipient invasions. It seems unlikely, however, that each federal agency would need to create its own separate team modeled after

those of NPS, because the duplication of manpower, equipment, and supplies could be substantial.

- *Measures to reduce the risk of exporting invasive species.* The United States might take further internal steps to avoid exporting potentially invasive species to other countries. These measures could be as simple as preventing their accidental export in bilateral aid programs or certifying that identified U.S. products (e.g., used tires) are free of pests. Such certification is done for agricultural shipments. Such a review might examine disaster aid and emergency relief, for example: in the rush to provide humanitarian relief, shipments of supplies, equipment, and personnel may inadvertently introduce diseases or pests unknown in the receiving country. Because such supplies sometimes are prepared for shipment in advance, they could be examined to reduce the risk of such transfers.[92] The NISC management plan considers international cooperation generally and describes actions that might be taken on a multilateral or bilateral basis to reduce the import and export of invasive species. However, it does not address steps the United States might take unilaterally, nor does it assess any positive or negative effects on U.S. trade that might occur from such steps. It is difficult to discern progress in this area in the last few years.

These options are not mutually exclusive. They likely would be under the jurisdiction of multiple committees in both House and Senate. They may offer opportunities for savings both to the economy and to ecosystems.

End Notes

[1] See the companion CRS Report R43258, Invasive Species: Major Laws and the Role of Selected Federal Agencieshttp://www.crs.gov/pages/Reports.aspx?PRODCODE=R43258 for available estimates.

[2] Also see Invasive Species Definition Clarification and Guidance White Paper Submitted by the Definitions Subcommittee of the Invasive Species Advisory Committee for a discussion of these terms. Available at http://invasivespecies.gov/ISAC/White%20Papers/ISAC_Definititions_White_Paper_FINAL_VERSION.pdf.

[3] For a general discussion of early development of federal wildlife law, see The Evolution of National Wildlife Law, 3rd ed., by Michael J. Bean and Melanie J. Rowland (Westport, CT: Praeger Publishers. 1997), pp. 7-14.

[4] For more information on the legal status of executive orders, see CRS Report RS20846, Executive Orders: Issuance, Modification, and Revocation.

[5] Invasive species are defined in §1 of Executive Order 13112 as "alien species whose introduction does or is likely to cause economic or environmental harm or harm to human health."

[6] CRS Report R43258, Invasive Species: Major Laws and the Role of Selected Federal Agencies. This report also covers the roles of a number of federal agencies, as well as providing statistics on economic impacts.

[7] Act of May 25, 1900, §2 and §4, 31 Stat. 188. For more on the provisions of the Lacey Act relating to injurious species, see CRS Report R43170, Injurious Species Listings Under the Lacey Act: A Legal Briefing.

[8] U.S. Congress, Office of Technology Assessment (OTA), Harmful Non-Indigenous Species in the United States, OTA-F-565 (Washington, DC: U.S. Government Printing Office, September 1993), p. 163.

[9] History did not provide grounds for optimism that the new order would be implemented effectively: the OTA Report noted (p. 166) that an older order (E.O. 11987, May 24, 1977) "in practice ... has been ignored by most Federal agencies. Moreover, the Fish and Wildlife Service has yet [15 years later] to implement the order in regulations although specifically directed to do so." Opposition to proposed Fish and Wildlife Service (FWS) regulations, developed pursuant to the Lacey Act, came primarily from agriculture, zoos, game ranches, aquaculture, and the pet trade. A revised version was opposed strongly by the hobby fish industry. The OTA report also noted that at that time no major industry supported the proposed FWS regulations. However, escapes of various species mentioned throughout this report, plus evidence of their economic impact, appears to have led to somewhat wider acceptance for regulation and control measures.

[10] This report has not been updated.

[11] A serious infestation of leafy spurge (Euphorbia esula) on an Oregon ranch caused a drop in property value from about $125-$150 per acre to $22 per acre over 10 years. (Federal Interagency Committee for the Management of Noxious and Exotic Weeds. Invasive Plants: Changing the Landscape of America. Washington, DC, p. 27; hereinafter referred to as the "FICMNEW report.")

[12] Although the transfer of human pathogens such as smallpox, syphilis, measles, AIDS, and malaria from continent to continent has enormous human, ecological, and economic consequences, this report will not cover human pathogens and parasites. It will, however, mention a few disease vectors (species such as mosquitos that can transmit diseases to other organisms but are not themselves pathogens).

[13] Non-native species are currently being investigated to sustain a nascent biofuels program. The search for a hardy plant that is able to grow vigorously on poor soils with low requirements for water and fertilizer is a search for the same characteristics that make plants invasive. To avoid the same pitfalls as kudzu, growers may need to cultivate sterile varieties of such species as miscanthus (a grass) or take other steps to prevent its unwanted spread.

[14] Measuring the number of non-native species that do no harm is exceedingly difficult: by being innocuous, these species escape study. For example, Kentucky bluegrass (Poa pratensis, which, despite its common name, may have been an early import from Europe) now serves as important forage for native herbivores in much of the United States, although its benefits are hard to quantify.

[15] Some authors call the phenomenon evolution in reverse, although this can be misleading because species themselves do not return to previous evolutionary forms. Rather ecosystems become more similar, contrary to typically increasing divergence. See Christopher Bright,

Life Out of Bounds: Bioinvasion in a Borderless World. New York, NY: W.W. Norton & Co, p. 17 (1998). (Hereinafter referred to as "Bright.")

[16] Near neighbors must be construed biologically. For example, a spiderling dispersed by wind might easily be blown from British Columbia to Montana. A freshwater clam or deep soil insect would scarcely ever travel so far naturally.

[17] J.T. Lisle, J.J. Smith, and D.D. Edwards, et al., "Occurrence of microbial indicators and Clostridium perfringens in wastewater, water column samples, sediments, drinking water, and Weddell seal feces collected at McMurdo Station, Antarctica," Applied Environmental Microbiology, vol. 70, no. 12 (December 2004), pp. 7269-7276.

[18] However, the statement does not apply to all ecosystems. For example, in the higher elevation forests of the Sierra Madre in Mexico and the very similar forests of the Rockies, plants and animals could evolve in the comparative isolation between these two areas and be transported only recently with greater links of trade and traffic. Thus, the flora and fauna of either disconnected area could pose a threat to the other while the more continuously connected desert species are much less likely to do so.

[19] Some scientists argue that evidence is lacking on how disturbance affects susceptibility to invasions. But most scientists generally accept the idea that severe depletion of an ecosystem's flora and fauna (e.g., through fire, storm, volcanic eruption, etc.) does offer significant opportunities for newly arrived species because the new arrivals face reduced competition in a disturbed habitat.

[20] Comprehensive surveys of invasive species are difficult to find. This list was compiled by the Animal Rights Foundation of Florida, http://www.animalrightsflorida.org/incidentlist.html.

[21] For more on available estimates of numbers of invasive species, see CRS Report R43258, Invasive Species: Major Laws and the Role of Selected Federal Agencies.

[22] See http://epa.gov/greatlakes/invasive/index.html for further information about invasive animals and plants in the Great Lakes area.

[23] However, an expert at NISC stated that "working with [the Aquatic Nuisance Species Task Force], we [the NISC] have been updating work on pathways from 2005 and 2007 to include a revised set of pathway diagrams that will include links to available guidance. The revisions to the diagrams are complete and we are now going through a process to update our website to enable this function on-line. This work may provide a useful structure for categorizing the pathways.... Our three main categories of pathways are 1) trade and living industry (e.g., landscaping plants, aquaculture, pets); 2) transportation (e.g., ballast water, hull fouling); and 3) infrastructure and resource management (e.g., introductions for erosion control, canals/waterways, waste disposal)." Stanley Burgiel, NISC, Dec. 1, 2014. Personal communication. The Aquatic Nuisance Species Task Force (ANSTF) is "an intergovernmental organization dedicated to preventing and controlling aquatic nuisance species, and implementing the Nonindigenous Aquatic Nuisance Prevention and Control Act (NANPCA) of 1990." See http://www.anstaskforce.gov/default.php for more information.

[24] According to one author, "By far the most ecologically disruptive sector of the pet industry is the aquarium trade.... Of those exotic fish species established in the United States that are completely foreign to the country, about 65 percent arrived through the aquarium trade." (Bright), pp. 162-163.

[25] According to one study, the United States legally imported 572,158 reptiles in 2010. Ariel H. Collis and Robert N. Fenili, The Modern U.S. Reptile Industry, Georgetown Economic Services, LLC. Economic Analysis Group, January 5, 2011. For an analysis of the illegal

wildlife trade, see CRS Report RL34395, International Illegal Trade in Wildlife: Threats and U.S. Policy.

[26] For an overview of the risks of both plants and animals introduced from the aquarium trade, see http://www.vetcentric.com/magazine/magazineArticle.cfm?ARTICLEID=1443.

[27] S.H. Reichard and C.W. Hamilton, "Predicting Invasions of Woody Plants Introduced into North America," Conservation Biology, vol. 11 (1997):193-203. For example, the giant hogweed, whose toxic sap can cause severe scarring, was introduced as an ornamental but escaped cultivation and has been listed for decades as a noxious weed. Mary Robson, "The Perils of Giant Hogweed," July 5, 1998, at http://gardening.wsu.edu/column/07-05-98.htm.

[28] FICMNEW report, pp. 86-91.

[29] After considerable controversy and scientific analysis about the potential effects of introducing the non-native Asian species, in 2009 the states of Maryland and Virginia rejected the proposed introduction. See http://chesapeakebay.noaa.gov/monitoring-and-research/non-native-oyster-research.

[30] For more on the role of the Animal and Plant Health Inspection Service (APHIS), see "Animal and Plant Health Inspection Service," below.

[31] For more on the unexpected consequences of the use of these beetles, see, for example, Bureau of Reclamation, Tamarisk Leaf Beetle Invasion Into The Colorado River Ecosystem Downstream From Glen Canyon Dam, at https://www.usbr.gov/uc/rm/amp/twg/mtgs/09sep29/Attach_15.pdf.

[32] The National Institute for Invasive Species Science (NIISS) website is at http://www.niiss.org/cwis438/websites/ niiss/Home.php?WebSiteID=1.

[33] The list of services provided offers no direction on how to obtain the services.

[34] Mark Williamson, Biological Invasions (London: Chapman & Hall, 1996), Chapter 2, pp. 28-54; and Daniel Simberloff, "The Role of Propagule Pressure in Biological Invasions," Annual Review of Ecology, Evolution, and Systematics, vol. 40 (2009), pp. 81-102.

[35] For a sample Hazard Analysis and Critical Control Points (HACCP) analysis (applied to the risk of transporting invasive species during the harvest of aquatic weeds), see http://seagrant.uconn.edu/whatwedo/ais/mechanical.doc.

[36] Only a few commercial foods now grown in the United States are native to this country. Known examples of native foods include sunflowers, Jerusalem artichokes, pecans, black walnuts, some wild cherries, strawberries, blueberries, and cranberries. Other familiar foods of the new world (e.g., corn, potatoes, chilies, and tomatoes) were non-natives brought north by Native Americans before 1492. For some foods, the native range is still debated.

[37] No estimates of the range of benefits of desirable non-native species were found.

[38] Guidelines For Assessing The Risk Of Non-Native Animals Becoming Invasive. November, 2011, at http://www.oie.int/fileadmin/Home/eng/Our_scientific_expertise/docs/pdf/OIE Guidelines_NonNativeAnimals_2012.p df.

[39] Non-native Species Manual. Committee on Environmental Protection, Antarctic Treaty Secretariat. 2011, at http://www.ats.aq/documents/atcm34/ww/atcm34_ww004_e.pdf.

[40] Code of Practice on Introductions and Transfers of Marine Organisms, 2005, at http://www.ices.dk/publications/Documents/Miscellaneous%20pubs/ICES%20Code%20of %20Practice.pdf.

[41] For an overview of plant quarantine procedures and guidelines, see Robert P. Kahn, Plant Protection and Quarantine. Vol. 1 Biological Concepts (Boca Raton, FL: CRC Press, Inc., 1989), p. 226.

[42] Identification of what might constitute such a manageable/tolerable level is subjective, value laden, and open to interpretation, depending upon who is affected.

[43] Early detection and rapid response can be defined as explicit and well-coordinated direction and management in the critical period between the first discovery (or intentional release) of a new non-native species and its establishment.

[44] In the absence of eradication, residual pest populations could otherwise reproduce and return to problem levels, so control actions may need to continue indefinitely.

[45] The degree of species specificity of the selected approach can be a valuable asset in targeting control efforts. In addition, success may be influenced by control program managers' efforts to solicit public input early in the process of formulating and evaluating control alternatives and to answer questions from the public about possible human health, economic, and other effects from control programs.

[46] University of California, Statewide Integrated Pest management Program (UC IPM), "Definition of Integrated Pest Management," at http://www.ipm.ucdavis.edu/ GENERAL/ipmdefinition.html.

[47] University of California, UC IPM, "What is Integrated Pest Management (IPM)," at http://www.ipm.ucdavis.edu/ GENERAL/whatisipm.html.

[48] Ibid.

[49] See, for example, Wisconsin Department of Natural Resources, "Best Management Practices (BMPs)," at http://dnr.wi.gov/topic/Invasives/bmp.html.

[50] U.S. Forest Service, Non-native Invasive Species Best Management Practices, August 2012.

[51] For examples, see Wisconsin Department of Natural Resources, "Prevention, Aquatic," at http://dnr.wi.gov/topic/Invasives/prevention.html.

[52] For example, the Environmental Protection Agency (EPA) approved methyl bromide as a fumigant for the brown tree snake. For more information, see CRS Report RS20863, Methyl Bromide and Stratospheric Ozone Depletion.

[53] However, care is required to ensure that barriers do not impede the migratory behavior or natural dispersal of native species.

[54] Refers to rotating crops grown in a field to disrupt the growth patterns of an animal or plant pest, rather than growing a single crop that may allow pest populations time to become established over time.

[55] Refers to spreading a clear plastic sheet over the soil, which traps solar heat that in turn kills soil-borne diseases, insects, nematodes, and many weed seeds.

[56] See, for example, North Carolina State University, "Pest Control Tactics," 2003; UC IPM, "What is Integrated Pest Management (IPM),"; Oklahoma Cooperative Extension, "Cultural Control Practices," HLA-6431; and Clemson University, "Controlling Weeds by Cultivating and Mulching," HGIC-1253.

[57] Traps have been used to control brown tree snakes in some areas of Guam, for example. See, for example, U.S. Department of the Interior, The Brown Tree Snake, Boiga irregularis, A Threat to Pacific Islands, Biological Report 88(31), (Washington, DC: Fish and Wildlife Service, September 1988), pp. 18-20.

[58] USDA requires proof of host specificity before supporting an insect introduction for non-native plant control.

[59] 7 C.F.R. 371.

[60] This section was originally prepared by Linda J. Schierow, specialist in Environmental Policy, now retired.

[61] UC IPM, "What is Integrated Pest Management (IPM)," at http://www.ipm.ucdavis.edu/ GENERAL/whatisipm.html; and UC IPM, "Applying IPM in Your Home and Landscape," at http://www.ipm.ucdavis.edu/GENERAL/ whatisipmurban.html.

[62] The use of methyl bromide is being phased out, except for certain critical uses. Among the critical-use exemptions are certain applications to control invasive species. See http://www.epa.gov/ozone/mbr/cueuses.html for recent exemptions that have been allowed.

[63] EPA, "What are Biopesticides?" http://www.epa.gov/pesticides/biopesticides/whatarebio pesticides.htm.

[64] In some species (particularly insects), chemicals known as sex pheromones are given off that allow males and females to find each other. Sex pheromones work even when target species densities are low, and they are highly species-specific. For example, Disparlure (the commercially synthesized sex pheromone of the female gypsy moth) is used to trap male gypsy moths. Because it is sometimes difficult to determine whether a natural pesticide controls a pest by nontoxic means, EPA established a committee to determine whether a pesticide meets the criteria for consideration as a biochemical pesticide.

[65] Public concern has arisen over the use of Bt in some situations (e.g., for controlling gypsy moths) due to concern for possible adverse effects on the human immune system. See I. L. Bernstein, J. A. Bernstein, and M. Miller, et al., "Immune responses in farm workers after exposure to Bacillus thuringiensis pesticides," Environmental Health Perspectives, vol. 107, no. 7 (July 1999), pp. 575-582.

[66] Both the genetic material and the protein are regulated by EPA; the plant itself is not regulated.

[67] For example, see several press reports, cited in http://www.kingsnake.com/blog/ archives/1120-Python-kill-totalstalls-around-50-as-Floridas-bounty-hunt-comes-to-an-end.html.

[68] State of New Jersey, Department of Agriculture. "New Jersey Declares Itself Free From Devastating Tree-Killing Pest," March 13, 2013, at http://www.nj.gov/agriculture/ news/press/2013/approved/press130314.html.

[69] Barbara Liston, "Bomb-sniffing dogs enlisted to stem Florida python invasion," Reuters, April 30, 2012.

[70] Kim M. Goodwin, Rick E. Engel, and David K. Weaver, "Trained Dogs Outperform Human Surveyors in the Detection of Rare Spotted Knapweed (Centaurea stoebe)," Invasive Plant Science and Management, vol. 3, no. 2 (April/June 2010), pp. 113-121.

[71] For more information, see CRS Report R43258, Invasive Species: Major Laws and the Role of Selected Federal Agencies.

[72] Control, rather than eradication, is nearly always the main option once a species is widely dispersed. The authors in the course of their research discovered very few instances (e.g., smallpox, nutria in East Anglia, UK) of control measures on a well-established species leading to long-term eradication. (For an example of a recent eradication, see "Rat Island is Officially Rat Free" at http://www.fws.gov/alaska/nwr/akmar/rat_island.htm; documenting an eradication project on a remote island in the Alaska Maritime National Wildlife Refuge. After the project, nesting seabirds recolonized the island, which was renamed Hawadax Island in 2012.) For established species, eradication is extremely improbable. For very recent arrivals, it could be possible. In contrast, prevention (here including a reduction in the number of non-natives establishing breeding populations each year) is probably more feasible.

[73] The roles, responsibilities, and statutory authorities currently exercised by federal agencies, including the four listed below, are described in the companion CRS Report R43258, Invasive Species: Major Laws and the Role of Selected Federal Agencies.

[74] The sale of some of these plants is forbidden by some states, but the plant could be purchased legally in one state and then moved to another, thereby obviating the protection of the

second state's laws. Some nurseries are taking steps to avoid the sale of invasive plants and/or to encourage the sale of native plants.

75 Although the noxious aquatic plant program has been active in the past, it has been reduced somewhat in a shift from federal to greater local responsibilities for maintenance and operation costs.

76 For the Corps' most recent strategy concerning Asian carp and its partnership with a number of local, state, federal, provincial and Canadian agencies, see Asian Carp Control Strategy Framework, by the Asian Carp Regional Coordinating Committee. June 2014.

77 Act of May 25, 1900, §2 and §4, 31 Stat. 188. For more on the provisions of the Lacey Act relating to injurious species, see CRS Report R43170, Injurious Species Listings Under the Lacey Act: A Legal Briefing.

78 P.L. 93-205, 87 Stat. 884. 16 U.S.C. §§1531-1544.

79 Department of the Interior, "National Environmental Policy Act: Implementing Procedures; Addition to Categorical Exclusions for U.S. Fish and Wildlife Service," 78 Federal Register 39307-39309, July 1, 2013.

80 Testimony of David Hoskins, Assistant Director for Fish and Aquatic Conservation, Fish and Wildlife Service, in House Committee on Natural Resources, Subcommittee on Fish, Wildlife, Oceans, and Insular Affairs, Oversight Hearing on "The Department of the Interior's Proposal to use a Categorical Exclusion Under the National Environmental Policy Act (NEPA) for Adding Species to the Lacey Act's List of Injurious Wildlife;" Sept. 20, 2013, Ser. No. 113-44, p. 16.

81 One method used to predict such risks is that of Hazard Analysis and Critical Control Points (HACCP), to create a stepped analysis of the critical points at which an invasion may be prevented or stopped. For a sample HACCP analysis (applied to the risk of transporting invasive species during the harvest of aquatic weeds), see http://seagrant.uconn.edu/whatwedo/ais/mechanical.doc.

82 P.L. 101-646 (104 Stat. 4761, 16 U.S.C. 4701); as amended by the National Invasive Species Act, P.L. 104-332.

83 Over the years, Congress has considered many issues related to particular invasive species. The list of options below focuses only on topics that cover a broad range of species, pathways, or agencies.

84 NISC, Meeting the Invasive Species Challenge, pp. 32-33.

85 Adam S. Davis, Roger D. Cousens, and Jason Hill, et al., "Screening bioenergy feedstock crops to mitigate invasion risk," Frontiers in Ecology and the Environment, vol. 8, no. 10 (December 2010), pp. 533-539.

86 While Smokey Bear has become widely recognized, other such efforts have been less successful. For example, the Forest Service's other symbol, Woodsey Owl, has been far less effective in reducing pollution and littering. In the early 1960s, USDA attempted to create "Pestina" to help educate the traveling public to the perils of non-native species imports. There was little effort to evaluate effectiveness, and little work with private and other federal agencies. The program was eventually dropped. (OTA Report, p. 142.) "Sammy Soil," however, has been an icon at the Natural Resources Conservation Service since 1967 and is featured in his own coloring book "The Adventures of Sammy Soil."

87 Appendix III of the Convention on International Trade in Endangered Species could provide a model for this type of approach. A nation may list any species native to it under this Appendix; through listing, it requests the assistance of other member nations in controlling imports of that species when they arrive from the host nation.

[88] Some local laws already make a similar distinction favoring sterilized animals when they charge a lower licensing fee for dogs that have been spayed or neutered than for intact dogs. The authors are unaware of any similar benefit for sterilized animals at the federal level.

[89] The Wild Bird Conservation Act (16 U.S.C. 4901 et seq.) could provide a model for legislation to encourage "homegrown" (if not native) species over additional foreign imports.

[90] The federally coordinated National Interagency Fire Center is coordinated primarily through the Bureau of Land Management. All federal land managing agencies participate, and a great deal of the program focuses on work with tribal, state, and local governments to bring many resources to bear on major fires.

[91] For more on Emergency Plant Management Teams, see http://www.nature.nps.gov/biology/invasivespecies/ EPMT_teams.cfm.

[92] In the case of introduced human diseases, a tragic example occurred after a major earthquake in Haiti in October 2010. Nepalese soldiers arriving under United Nations auspices apparently inadvertently introduced cholera (Vibrio cholerae) into an already catastrophic situation. Cholera had been unknown in Haiti for many decades before the earthquake. See U.N. Report: Final Report of the Independent Panel of Experts on the Cholera Outbreak in Haiti, by Alejandro Cravioto, et al.

In: Invasive Species Management
Editor: Jeannie Saunders

ISBN: 978-1-63485-064-3

Chapter 2

INVASIVE SPECIES: MAJOR LAWS AND THE ROLE OF SELECTED FEDERAL AGENCIES*

M. Lynne Corn and Renée Johnson

SUMMARY

An "invasive" species (alternatively known as an alien, exotic, injurious, introduced or naturalized, non-native, nonindigenous, nuisance, or noxious species) refers to an animal or plant that is introduced into an environment where it is not native. The introduction of invasive species to the United States—whether deliberate or unintentional—from around the globe can pose a significant threat to native animal and plant communities, and may result in extinctions of native animals and plants, species disruptions as native and non-native species compete for limited resources, reduced biodiversity, and altered terrestrial or aquatic habitats. This can result in a range of economic, ecologic, and cultural losses, including reduced agricultural output from U.S. farms and ranches; degradation of U.S. waterways, coastal areas, national parks, and forests; and altered urban, suburban, and rural landscapes.

It is estimated that 50,000 non-native species have been introduced to the United States. The potential economic costs associated with nonindigenous plant and animal species are estimated at $129 billion annually in the United States. A few examples of the types of damages attributed to non-native

* This is an edited, reformatted and augmented version of a Congressional Research Service Publication, CRS Report for Congress R43258, dated October 24, 2013.

invasive species in the United States are as follows. Burmese pythons are multiplying in south Florida, becoming a top carnivore and killing large numbers of native species of reptiles, birds, and mammals. Zebra and quagga mussels from Eastern Europe are clogging intakes for urban water supplies and nuclear power plants in the Great Lakes and the Mississippi basin. The light brown apple moth, a native pest of Australia, has been detected in California and is causing damage to a wide range of plant species and commercial fruit and vegetable crops. Leafy spurge is lowering the forage value of western grazing land, and reducing overall land values.

In the United States, numerous federal and interagency efforts share responsibilities regarding invasive species. Among the federal agencies involved are the Departments of Agriculture, Commerce, Defense, Homeland Security, Interior, Transportation, and others, including the Environmental Protection Agency and the Executive Office of the President. Of these, three Departments—Agriculture, Commerce, and Interior—play a major role by co-chairing the National Invasive Species Council (NISC). Created by Executive Order 13112 in 1999, NISC provides high-level interdepartmental coordination of federal invasive species actions and works with other federal and nonfederal groups to address invasive species issues at the national level.

In FY2012, the U.S. government spent an estimated $2.2 billion across a range of federal agencies and activities in an effort to prevent, control, and eradicate invasive species domestically. Activities at the Department of Agriculture accounted for the bulk of available federal funding, nearly $1.3 billion (58% of total available funds). Activities at the Department of Homeland Security, comprised of mostly border protection and security activities, accounted for about $0.7 billion (31% of total funding). The remainder of federal funding, about $0.2 billion (about 11% of total funding) covers activities across a range of agencies at the Departments of Interior, Commerce, and Defense, and also other independent agencies.

Despite efforts to achieve high-level interdepartmental coordination, comprehensive legislation on the treatment of invasive species has never been enacted, and no single law provides coordination among federal agencies. Instead, the current legal framework is largely governed by a patchwork of laws, regulations, policies, and programs. Some laws are tailored to individual species or narrowly focused on what is affected by the species. Other laws have a broader intended purpose and may only peripherally address invasive species. Some laws, although they do not directly address invasive species control or prevention, may limit such introductions.

An "invasive" species (alternatively known as an alien, exotic, injurious, introduced or naturalized, non-native, nonindigenous, nuisance, or noxious species) refers to an animal or plant that is introduced into an environment where it is not native.[1] The introduction of invasive species to the United States—whether deliberate or unintentional—from around the globe can pose a significant threat to native animal and plant communities, and may result in extinctions of native animals and plants, species disruptions as native species compete for limited resources, reduced biodiversity, and altered terrestrial or aquatic habitats.

It is estimated that 50,000 non-native species have been introduced to the United States, including nonindigenous plant and animal species (**Table 1**).[2] Examples of some invasive species found in the United States, and the types of damages they may inflict, include the following:

- Burmese pythons (*Python bivitattus*) have multiplied in south Florida, becoming a top carnivore and killing large numbers of native species of reptiles, birds, and mammals.
- Brown tree snakes (*Boiga irregularis*) from the western Pacific continue to threaten power utilities and communications on Guam and the Northern Marianas, and threaten to invade Hawaii and harm its tourism industry.
- Zebra mussels (*Dreissena polymorpha*) and quagga mussels (*Dreissena rostriformis bugensis*) from Eastern Europe have clogged intakes for urban water supplies and nuclear power plants in the Great Lakes and the Mississippi basin.
- Light brown apple moth (*Epiphyas postvittana*), a native pest of Australia, has been detected in California and is causing damage to a wide range of plant species and commercial fruit and vegetable crops.
- Lionfish (*Pterois volitans*), likely released from private aquaria in Florida, have spread north along the U.S. coast and also within the Caribbean.
- Formosan termites (*Coptotermes formosanus*) have devastated living trees and historic buildings in the French Quarter of New Orleans.
- Leafy spurge (*Euphorbia esula*) has reduced the forage value of western grazing land, resulting in lower overall value to private landowners.

All 50 states and the U.S. territories have at least some invasive plants and animals. A few (e.g., Hawaii, Florida, Louisiana, the Great Lakes states, and

California) have so many harmful non- natives as to cause major ecological and economic damage to a variety of locations and industries.

This report provides an overview of the federal laws and directives in the United States that govern invasive species, and the role of selected federal agencies. A summary of selected laws and agencies is provided in **Appendix A**. The patchwork of laws that currently govern invasive species in the United States also contribute to fundamentally different approaches to regulate invasive species, which is beyond the scope of this report.

In addition to the federal laws discussed in this report, invasive species are regulated and/or managed through a variety of statutes administered by the states, often with the cooperation of federal agencies. This report also does not address the extensive efforts of individual states to deal with invasive species concerns.[3]

Further, this report focuses primarily on invasive plant and animal species that may affect U.S. agriculture, waterways and coastal areas, national parks and forests, and various ecosystems and landscape environments. The potential for non-native species to adversely affect human health and disease is not addressed in this report.

This report also does not address species introductions to the United States that are widely considered beneficial, such as certain crop plants (e.g., wheat, soybeans) and orchard fruit (e.g., apples, pears), as well as honey bees (*Apis mellifera*) and pheasants (*Phasianus colchicus*), among other introduced species.

ESTIMATED ECONOMIC COSTS

The introduction of invasive animal and plant species can result in a range of economic, ecologic, and cultural losses, including but not limited to reduced agricultural output from U.S. farms and ranches, degradation of U.S. waterways and coastal areas, as well as national parks and forests, and changed urban, suburban, and rural landscapes.

It is difficult to quantify and further monetize the economic damage due to a non-native invasive species. The most widely cited and available aggregation of the various available cost estimates is by researchers at Cornell University.[4] The most recent reported estimates put the potential aggregate economic costs from selected invasive species at $127 billion annually in the United States (**Table 1**), excluding estimated costs for human diseases (which are not covered in this report).[5] These estimated economic costs include control costs and direct damages and losses to property values, agricultural or natural resource productivity or output, and other costs.

Based on these estimated economic costs, the single largest cost is damage to U.S. agricultural crop and livestock production, totaling nearly $65 billion annually. This estimate is for losses, damages, and control costs associated with agricultural crop weeds (estimated at $20.5 billion annually); crop plant pathogens ($18.4 billion); livestock diseases ($9 billion); and weeds in pastures ($6 billion). Plant pathogens and pests in forests account for another $4.2 billion annually. Other categories include rats ($19 billion), fish ($5.4 billion), and weeds and plant pathogens in lawns, gardens, and golf courses (totaling $3 billion annually).

The text box below highlights available individual estimates of the estimated costs of some selected invasive species and describes their potential impacts.

These cost estimates do not account for possible offsetting benefits from some introduced species. Some criticize the estimates for overstating the cost of damages from invasive species, or for not considering possible offsetting benefits.[6] The estimates also do not account for some types of indirect losses and costs, or certain nonmarket welfare losses. Others note that a full accounting would consider other types of losses and would result in a higher estimate.[7]

Estimated Economic Costs of Selected Invasive Species

The following are individual estimates by various sources of the costs associated with selected invasive species.

- Leafy spurge (*Euphorbia esula*): Costs an estimated $100 million annually in cattle forage losses in North Dakota, but spreading throughout the northern Great Plains and Rocky Mountain areas.
- Purple loosestrife (*Lythrum salicaria*): Costs an estimated $45 million annually across nearly all U.S. states, attributable to forage losses and to control costs.
- Sea lamprey (*Petromyzon marinus*): Costs an estimated $680 million annually from recreational fishery losses and control costs in the northern United States and Canada.

- Zebra mussel (*Dreissena polymorpha*): Costs an estimated $1 billion annually in recreational fishery losses, controls, and other costs in the Great Lakes and Pacific Northwest areas, as well as cleaning of water intake pipes, filtration equipment, power generating equipment, damage to docks and recreational or commercial boats.
- Hydrilla (*Hydrilla verticillata*): Costs an estimated $860 million in Florida in damages to agriculture, flood control, and residential property values.
- Formosan termites (*Coptotermes formosanus*): Costs an estimated $1 billion annually (including several hundred million dollars in New Orleans alone).
- Imported fire ants (*Solenopsis invicta and S. ricteria*): Costs an estimated $1 billion annually (including $300 million in Texas alone).
- Brown tree snakes (*Boiga irregularis*): Costs an estimated $12 million annually (in Guam alone, management and controls cost $4 million per year; other unaccounted for damages include power outages, slowed transportation and shipping, lost agricultural productivity and recreation/tourism, and healthcare costs from snake bites; other unaccounted for costs include the loss of bird species attributable to the snake in Guam).
- Weeds, pests, and plant and animal diseases affecting U.S. agricultural crop and livestock production total nearly $65 billion annually.

Sources: Estimates are from D. Pimentel, "Environmental and Economic Costs Associated with Alien Invasive Species in the United States," *in Biological Invasions: Economic and Environmental Costs of Alien Plant, Animal, and Microbe Species*, CRC Press, 2011; S. Lovell et al., "The Economic Impacts of Aquatic Invasive Species: A Review of the Literature," *Agricultural and Resource Economics Review*, Vol. 35(1):195-208, April 2006; L. Sandell and S. Knezevic, "Leafy Spurge," University of Nebraska, http://www.ianrpubs.unl.edu/live/ec174/build/ec174.pdf; and Federal Interagency Committee for the Management of Noxious and Exotic Weeds (FICMNEW), *Invasive Plants: Changing the Landscape of America*, 2003.

Table 1. Estimated Annual Costs Associated with Some Non-Native Species, Introductions in the United States

Category	Non-Indigenous Species	Losses and Damages	Control Costs	Total ($million)
	(number)	($ million)	($ million)	($ million)
Plants	**25,000**			
Purple loosestrife		–	–	45
Aquatic weeds		10	100	110
Melaleuca tree		na	3-6	3-6
Crop weeds		17,500	3,000	20,500
Weeds in pastures		1,000	5,000	6,000
Weeds in lawns, gardens, golf courses		na	1,500	1,500
Subtotal				**28,155**
Mammals	**20**			
Wild horses, burros		5	na	5
Feral pigs		1,000	<1	1,000
Mongooses		50	na	50
Rats		19,000	na	19,000
Cats		18,000	na	18,000
Dogs		425	na	425
Subtotal				**38,481**
Birds	**97**			
Pigeons		2,200	na	2,200
Starlings		800	na	800
Subtotal				**3,000**
Reptiles, Amphibians	**53**			
Brown tree snake		3	14	16
Subtotal				**16**
Fish	**138**	**5,400**	**na**	**5,400**
Arthopods	**4,500**			
Imported fire ant		1,200	800	2,000
Formosan termite		1,000	na	1,000
Green crab		44	na	44
Gypsy moth		na	11	11
Crop pests		10,400	500	10,900
Pests in lawns, gardens, golf courses		na	1,500	1,500
Forest pests		2,100	na	2,100
Subtotal				**17,555**
Mollusks	**88**			

Table 1. (Continued)

Category	Non-Indigenous Species (number)	Losses and Damages ($ million)	Control Costs ($ million)	Total ($million) ($ million)
Zebra mussel		–	–	2,000
Asian clam		1,000	na	1,000
Shipworm		205	na	205
Subtotal				**3,205**
Microbes	**20,000**			
Crop plant pathogens		18,000	400	18,400
Plant pathogens in lawns, gardens, golf courses		NA	2,000	2,000
Forest plant pathogens		2,100	na	2,100
Dutch elm disease		na	100	100
Subtotal				**22,600**
Livestock Diseases		**9,000**	**na**	**9,000**
TOTAL (excluding human disease)				**$127,418[a]**

Source: D. Pimentel, "Environmental and Economic Costs Associated with Alien Invasive Species in the United States," in *Biological Invasions: Economic and Environmental Costs of Alien Plant, Animal, and Microbe Species*, CRC Press, 2011. May not add due to rounding. na = "not available."

Notes: Updates previous estimates from the first edition of the book, as well as estimates from previous studies in 2000 and 2005: D. Pimentel et al., "Update on the Environmental and Economic Costs Associated with Alien- Invasive Species in the United States," *Ecological Economics*, February 2005, Vol. 52(3): 273-288; and D. Pimentel et al., "Environmental and Economic Costs of Nonindigenous Species in the United States," *BioScience*, January 2000, Vol. 50: 53-65.

[a] Excludes human diseases, estimated at nearly $92 billion annually for reportedly introduced diseases such as AIDS, syphilis, and influenza. This report focuses on invasive plant and animal species that may affect U.S. agriculture, waterways and coastal areas, national parks and forests, and various ecosystems and landscape environments, rather than human health and disease.

SELECTED FEDERAL LAWS AND DIRECTIVES

Comprehensive legislation on the treatment of invasive species has never been enacted, and no single law directs coordination among federal agencies. No laws focus on the broad problems of invasive species, their interception, prevention, and control across a variety of industries and habitats. Instead, the current legal framework is largely governed by a patchwork of laws, regulations, policies, and programs. Some laws are tailored to individual species or narrowly focused on what is affected by the species, such as agricultural production or certain aquatic or terrestrial ecosystems. Other laws have a broader intended purpose and may only peripherally address invasive species, such as certain environmental laws, resource management laws, and species or wildlife protection laws. Some laws, though they do not directly address invasive non- native species control or prevention, have effects that may limit such introductions. Below is a brief digest of existing laws that affect non-native species introduction, prevention, and control. However, control of invasive species is not often the major purpose of the law in some cases, and agencies have little authority to eradicate invasive species, except where they occur on federally managed lands, thus undermining the effectiveness of these programs. In general, laws addressing threats to agriculture (for centuries a well-developed North American industry whose risks from non-native invasion species are relatively clear) tend to be more developed than laws protecting other industries or ecosystems.

This patchwork contributes to two fundamentally different regulatory approaches to address invasive species: one approach based on a particular listed or known species, and another approach based on certain pathways by which a variety of species may be introduced (see text box). Further discussion of these two regulatory approaches is beyond the scope of this report.

Approaches to Invasive Species Regulation: Single Species vs. Pathways

Single-Species Approach

Under a single-species approach, regulation of invasive plants or animals must be placed on a "black list" before they are regulated as harmful. Black list approaches to invasive species are, of necessity, done on a species-by-species basis. Harm can rarely be demonstrated unless the plant or animal is already at pest levels and inflicting damage somewhere—that is, generally after the species is reproducing and spreading. Usually, damage must be readily apparent before protection can begin, at which point prevention could be nearly impossible. A key factor is knowledge of the presence of the species.

If the species, its potential damage, or its means of transport are unknown, it will not be regulated under the black list approach. Examples of this approach are demonstrated in the laws described in this report. One such example includes coverage of the brown tree snake under the Nonindigenous Aquatic Nuisance Prevention and Control Act of 1990 (NANPCA; 16 U.S.C. §§4701, *et seq.*), as amended by the National Invasive Species Act (NISA, 16 U.S.C. §§4701 *et seq.*), which otherwise regulates the pathway of ship ballast water. Another example is the requirement that individual weed species be placed on an exclusion list before they can be regulated under the Plant Protection Act of 2000 (PPA; 7 U.S.C. §§7701 *et seq.*).

Pathways Approach

Under a pathways approach, regulation of invasive species tends to be based on the risk of invasion via certain pathways. In such an approach, plants or animals may be removed wholesale, e.g., by sterilizing pallet wood, cleaning a cargo hold, or sterilizing the soil in which horticultural specimens are shipped. There are instances of regulation and prevention by pathway. Examples of this approach are demonstrated in the laws described in this report. One such example is NANPCA's regulation of ballast water as a risk to saltwater and freshwater ports, bays, and estuaries. Similarly, plant and animal quarantine and inspection requirements under both the PPA and the Animal Health Protection Act of 2002 (AHPA, 7 U.S.C. §§8301-8302) give the U.S. Department of Agriculture (USDA) broad authority to inspect imported agricultural products to detect, control, or eradicate plant and animal pests or diseases. Pathway approaches also exist at the state level. Pathway approaches do not require lists of organisms to be implemented or effective, and may even block the entry of species whose existence is unknown to science.

Federal Laws

The laws described here are listed according to the year the law was enacted. **Appendix A** provides a summary of the laws and statutes governing invasive species, and is based in part on information compiled by the National Invasive Species Council (NISC).[8]

This report does not provide a comprehensive review of all the U.S. laws and statutes that may contain provisions that address some aspect of invasive or non-native species. Some environmental laws (such as the Federal Insecticide, Fungicide, and Rodenticide Act and the Clean Water Act),[9] and certain resource management laws (such as the Fish and Wildlife Act and other laws)[10] are not directly covered in this report, although some provisions in these and other U.S. laws may play a role in addressing invasive species. Also omitted are federal laws referring to single species. This report does, however, cover the National Environmental Policy Act (NEPA), requiring federal agencies to take into consideration potential environmental impacts, including invasive species, from any planned agency actions.[11]

In addition to federal laws, a number of states have laws restricting transport or possession of invasive species. State laws are not described in this report.

Organic Administration Act

The Organic Administration Act of 1897 (16 U.S.C. §551) provides broad authority to the U.S. Forest Service within the U.S. Department of Agriculture (USDA) to protect National Forest System lands from a range of threats, including invasive species. In addition, under the Multiple- Use Sustained-Yield Act of 1960 (16 U.S.C. §§528-531), USDA manages U.S. national forests for multiple uses—such as outdoor recreation, range, timber, watershed, and wildlife and fish purposes.

Lacey Act[12]

The Lacey Act of 1900 (18 U.S.C. §§42-43; 16 U.S.C. §§3371-3378) addresses illegal wildlife trade to protect species at risk and bars importing species found to be injurious to the United States.[13] The portion of the Lacey Act known as the *injurious species provision* is codified in the criminal code at 18 U.S.C. §42.[14] The injurious species provision (18 U.S.C. §42) bans import and shipment of listed living creatures and their eggs.[15]

Under 18 U.S.C. §42(a)(1)), the Secretary of the Interior and the Secretary of the Treasury are may exclude the importation and shipment of three major categories of non-native animals: vertebrates, crustaceans, and mollusks. The list of just these three taxonomic categories means that other abundant and diverse groups of animals, such as insects and spiders, are not covered in the act's injurious species provisions.[16] Moreover, grounds for excluding such imports or shipments go beyond the traditional harm to agriculture, horticulture, and forestry interests to include harm to "wildlife or the wildlife resources of the United States." The act's broad definition of harm could mean that nearly any non-native vertebrate, crustacean, or mollusk could be considered for exclusion, because most and perhaps all ecologists would hold that the proliferation of any non-native species in an ecosystem risks harm to the nation's wildlife or wildlife resources.

The list of banned species may be amended either by statute or by regulation issued by FWS. Permits may be issued to import banned species for scientific, zoological, educational, or medical purposes. Under the injurious species provision, it is also illegal to import or ship between states any species listed under the act. A violation is a Class B misdemeanor, punishable by no more than six months in jail and/or up to a $5,000 fine for an individual and $10,000 for an organization. The species listed as injurious wildlife under the Lacey Act is available at the Department of Interior's U.S. Fish and Wildlife Service (FWS) website.[17]

Generally, laws that list banned species are known as "black list" laws. "White list" laws ban importing all species *except* those on an approved list.[18] For decades, the Lacey Act was primarily a white list law—prohibiting importing "any foreign wild animal or bird" except under special permit, as well as originally banning all imports of four species: mongoose, fruit bats, sparrows, and starlings.[19] The 1949 amendments of the Lacey Act transformed the injurious species provision into a purely black list law, such that only listed species were banned.[20] As a result of the black list approach, a species that might merit exclusion is not covered under the Lacey Act until a potentially lengthy review process is completed, thus limiting the usefulness of this statute when a new potential invader is first discovered.

Because the Lacey Act relies on a black list, it implicitly focuses on those species which are knowingly moved between states or nations, or at least travel on pathways already known to present a high risk of transporting unwanted plants or animals. If someone enters the United States, or crosses between states, unaware that the plant or animal has stowed away in a hubcap, on a wheel well, or beneath a shoe, and that plant or animal furthermore is not already on a black list, the injurious species provision of the Lacey Act has little or no bearing on the act.

Another provision of the statute is not discussed in detail here but also has implications for invasive species by making it illegal to import, export, transport, sell, receive, acquire, or purchase in interstate or foreign commerce any fish, wildlife, or plant taken, possessed, transported, or sold in violation of any federal, tribal, state, or foreign law (16 U.S.C. §3372 (a)(1), (2), and (4)).[21]

Virus-Serum-Toxin Act

The Virus-Serum-Toxin Act, as amended (21 U.S.C. §151 *et seq.*), was originally enacted in 1913 and authorizes USDA to regulate veterinary biological products that are intended for use in the treatment (i.e., prevention, diagnosis, management, or cure) of animal diseases. These include vaccines, bacterins, sera, antisera, antitoxins, toxoids, allergens, diagnostic antigens prepared from, derived from, or prepared with microorganisms, animal tissues, animal fluids, or other substances of natural or synthetic origin. The law prohibits the shipment or delivery for shipment in interstate and intrastate commerce, as well as the importation or exportation of any veterinary biological product that is "worthless, contaminated, dangerous, or harmful," and also any biological product not prepared in compliance with USDA regulations at a USDA-licensed establishment.[22]

Activities under the law are generally administered by USDA's Animal Plant Health Inspection Service (APHIS); however, there is an existing memorandum of understanding between APHIS and the Food and Drug Administration (FDA) at the Department of Health and Human Services (HHS) regarding procedures and responsibilities for resolving jurisdictional issues and questions concerning the regulation of certain animal products as biologicals under VSTA, or as drugs under the Federal Food, Drug, and Cosmetic Act (21 U.S.C. §321(g)(1)).[23]

Animal Damage Control Act

The Animal Damage Control Act of 1931, as amended (7 U.S.C. §§426 *et seq.*), is the primary statute under which USDA operates its Wildlife Services program.[24] This is the lead USDA program to conduct research and control work on invasive species to agriculture such as the brown tree snake, as well as address damage problems caused by such invasive species as nutria (*Myocastor coypus*), European starlings (*Sturnus vulgaris*), and monk parakeets (*Myiopsitta monarchus*). The law gives APHIS wide authority to control damage caused by wildlife to agricultural interests, including livestock, on federal, state, or private land. The program aims to protect field crops, vegetables, fruits, nuts, horticultural crops, and commercial forests; freshwater aquaculture ponds and marine species cultivation areas; livestock on public and private rangeland and in feedlots; public and private buildings and facilities, such as houses, commercial properties, swimming pools, golf courses, reservoirs, levees, and landfills; civilian and military aircraft (against collisions with birds); and public health (against wildlife-borne diseases such as rabies, Lyme disease, West Nile virus, and plague). Control methods include providing advice to individuals and to municipal, state or federal agencies on a wide variety of preventive, non-lethal control methods. Control of predatory animals, native or non-native, is largely carried out by lethal means, including hunting, trapping, and poisoning. The agency publishes annual Program Data Reports to inform the public on its wildlife damage management activities.[25]

APHIS has memoranda of understanding and other cooperative agreements with FWS, the National Park Service, the Bureau of Land Management, the Forest Service, and state natural resource agencies to help protect natural resources, including wildlife and threatened or endangered species, from loss of life, habitat, or food supply due to the activities of other species, including invasive species.

Soil Conservation and Domestic Allotment Act

The Soil Conservation and Domestic Allotment Act of 1936 (16 U.S.C. §590(a)-590(f)) gives USDA's Natural Resources Conservation Service (NRCS) the authority to operate Plant Materials Centers for the development, testing, and distribution of plants and vegetation management technologies for voluntary use by land owners and users of private or other nonfederal lands for soil erosion control, water conservation, and wildlife habitat. In addition, the NRCS Conservation Technical Assistance Program provides technical assistance to land owners and users of private or other nonfederal lands to plan and install, on a voluntary basis, structures and land management practices for soil erosion control and water conservation. These programs broadly provide for the detection and prevention of invasive species. Other USDA farmland conservation programs— such as the Environmental Quality Incentives Program, Wetlands Reserve Program, and Wildlife Habitat Incentives Program[26]—also provide technical, educational, and financial assistance to livestock and crop producers to protect against threats to soil, water, and related natural resources, and may also address invasive species concerns. Aspects of each these efforts encompass prevention, control and management, and restoration relating to invasive species.[27]

Federal Seed Act

The Federal Seed Act of 1939, as amended (7 U.S.C. §§1551 *et seq.*), requires accurate labeling and purity standards for seeds in commerce, and prohibits the importation and movement of adulterated or misbranded seeds. The law also authorizes enforcement activities and rulemaking functions. In addition, it regulates interstate and foreign commerce in seeds, and addresses "noxious weed seeds" that may be present in agricultural (e.g., lawn, pasture) or vegetable seed. APHIS administers the foreign commerce provision of this law. USDA's Agricultural Marketing Service administers the interstate commerce provisions. The law works in conjunction with the Plant Protection Act (7 U.S.C. §§7701 *et seq.*), which authorizes APHIS to regulate imports of agricultural seed when they may contain noxious weed seeds.

National Environmental Policy Act[28]

The National Environmental Policy Act of 1970 (NEPA, 42 U.S.C. §§4321 et seq.), as amended, established a national policy to protect the environment.[29] Federal agencies are required to comply with NEPA and consider the environmental impacts, including invasive species, of an agency's actions.

NEPA has two primary aims—to require federal agencies to consider the environmental effects of their actions *before* proceeding with them; and to involve the public in the decision-making process. To ensure that environmental impacts are integrated into that process, federal agencies must prepare an environmental impact statement for actions "significantly" affecting the quality of the human environment.[30]

NEPA applies only to "federal actions," defined broadly to include projects and programs entirely or partly financed, assisted, conducted, regulated, or approved by federal agencies.[31] Accordingly, programs or projects intended to control invasive species (e.g., BLM development of its Weed Management and Invasive Species Program), or actions that may result in the spread or introduction of non-native invasive species (among a range of other potential impacts), may be subject to NEPA. Such impacts may occur as a direct result of the action (e.g. state transportation agency landscaping or erosion control projects that receive federal funds) or be incidental to the action (e.g., a federally authorized construction project that opens a corridor that provides an opportunity for the movement of invasive species; or that introduces seeds from noxious weeds on construction equipment).

NEPA does not prohibit an agency from moving forward with a program or project that may introduce or spread non-native invasive species. Nor does NEPA require an agency to implement measures to control such impacts. Within the framework of completing the NEPA process, an agency would identify any environmental requirements applicable to a proposed action,[32] including any measures that must be taken to assure or demonstrate compliance with those requirements. To demonstrate compliance with those requirements, the NEPA analysis must document any outside agency review or consultation regarding the proposal, and identify any measures necessary to control, minimize, or mitigate regulated impacts.

Endangered Species Act

The Endangered Species Act of 1973, as amended (ESA, 16 U.S.C. §§1531-1543) focuses on the conservation and protection of endangered or threatened species and their habitats, except for species that are common to the point of being weeds or pests.[33] Although ESA has no direct regulation of invasive species, it could limit actions involving an invasive species to the extent the action may harm a listed species. For example, in the Pacific Northwest, the threat to resident Pacific salmon (*Onchorhynchus sp.*) protected under ESA is a major argument being used against the introduction or expansion of aquaculture that might introduce Atlantic salmon (*Salmo salar*), potentially an invasive species. Similarly, introduction of mountain goats (*Oreamnos americanus*) in an area where they are not native could be subject to proposed mitigation if the introduction would threaten listed plants likely to be eaten by the goats. ESA is jointly administered by the Departments of the Interior and Commerce.

ESA could provide protection in two ways. First, if the introduction were to be carried out by a federal agency or to require licensing, financial support, permits, or other support from a federal agency, the agency involved would have to consult with FWS or National Marine Fisheries Service (NMFS) to determine whether the introduction (or action leading to introduction) would tend to jeopardize the continued existence of the listed species or adversely modify its critical habitat.[34] If the agency action would lead to jeopardy or adverse modification, the action agency would need to carry out a reasonable and prudent alternative to avoid such problems, or risk violating the ESA. The alternative might, for example, reject the introduction in favor of a native species.[35] Second, if the action had no federal nexus, but its effects could result in taking (as defined in the act) a listed species, the party carrying out the action would have to obtain an incidental take permit from FWS or NMFS.[36]

Federal Noxious Weed Act

Most provisions in the Federal Noxious Weed Act of 1974 were supplanted by the Plant Protection Act;[37] however, a key section (7 U.S.C. §2814) still requires each federal agency to provide for noxious weed management on lands under its jurisdiction. The provision, introduced in the 1990 farm bill,[38] amended the Federal Noxious Weed Act to require federal agencies to establish and fund noxious weed management programs. It also allows the agencies to implement cooperative agreements with state agencies regarding the management of undesirable plant species in areas adjacent to federal lands. The law requires joint leadership from the Secretaries of Agriculture and the Interior in coordinating federal agency programs for control, research, and education associated with designated noxious weeds. In 1994, a memorandum of understanding among several federal agencies created the Federal Interagency Committee for Management of Noxious and Exotic Weeds (FICMNEW) as a vehicle to coordinate noxious weed priorities.[39]

Forest and Rangeland Renewable Resources Planning Act

The Forest and Rangeland Renewable Resources Planning Act of 1974 (16 U.S.C. §§1671 *et seq.*), as amended by the National Forest Management Act (16 U.S.C. §1604), is the U.S. Forest Service's primary authority to conduct research activities, including research relating to invasive species. The law contains broad authority for research and technology regarding U.S. lands related to the protection, conservation, and sustainable use of natural resources. The law also authorizes competitive grants to conduct research, and authorizes cooperative agreements with university, industry, and other private and public partnerships.

Federal Land Policy and Management Act

Provisions under the Federal Land Policy and Management Act of 1976, as amended (43 U.S.C. §1701 et seq.), provide funds for range betterment within a variety of range rehabilitation and improvement activities, including weed control on certain National Forest System rangelands.[40] In addition, the Public Rangelands Improvement Act of 1978 (43 U.S.C. §§1901-1908) provides funding for on-the-ground rangeland rehabilitation and range improvements on some of the rangelands managed by the Forest Service within USDA.

Cooperative Forestry Assistance Act

Under the Cooperative Forestry Assistance Act of 1978, as amended (16 U.S.C. §§2101-2111), USDA's Forest Service may enter into cooperative agreements to assist other federal, state, and private entities in controlling and managing invasive species on other federal lands and non- federal lands. The primary cooperative authority for invasive species is Section 8 of the law (16 U.S.C. §2104) and authorizes USDA to conduct activities and provide technical assistance relating to insect infestations and disease conditions affecting trees on National Forest System lands and on other federal lands (in cooperation with other federal agencies).[41] The law also provides support for good forest management practices, including financial assistance to maintain healthy timber ecosystem to prevent incursion of invasive species, on privately owned non- industrial forestlands.

International Forestry Cooperation Act

Provisions under sections of the International Forestry Cooperation Act of 1990 (16 U.S.C. §4501(b)) allow USDA to support international forestry and related natural resource activities and provide assistance to prevent and control insects, diseases, and other damaging agents, including invasive species. USDA's Forest Service delivers research and development to conduct prevention, rapid response, control, and management activities related to invasive species and to restore areas affected by invasive species.

Nonindigenous Aquatic Nuisance Prevention and Control Act

The Nonindigenous Aquatic Nuisance Prevention and Control Act of 1990 (NANPCA; 16 U.S.C. §§4701, *et seq.*) established a federal program to prevent the introduction of, and to control the spread of, unintentionally introduced aquatic nuisance species, including zebra mussels, and also the brown tree snake. NANPCA mandated a Great Lakes ballast water management program to prevent the introduction and spread of aquatic nuisance species into the Great Lakes through the ballast water of vessels and established civil and criminal penalties for violating these requirements.[42] Under the program, all ships entering U.S. waters (after operating outside the U.S. Exclusive Economic Zone) are directed to undertake high seas ballast exchange or alternative measures pre-approved by the Coast Guard as equally or more effective.

NANPCA established the Aquatic Nuisance Species Task Force (ANSTF), co-chaired by FWS and the National Oceanic and Atmospheric Administration (NOAA), which is tasked with implementing NANPCA.[43] NANPCA is the major legal authority for NOAA activities, and also covers some FWS activities.[44] Provisions in NANPCA primarily address prevention, control and management, and research of invasive species, but not rapid response and restoration activities.[45] NANPCA authorizes state governors to submit comprehensive management plans to ANSTF that identify areas or activities that need technical and financial assistance.[46] NANPCA encourages the Secretary of Transportation, through the International Maritime Organization, to negotiate with foreign countries on the prevention and control of the unintentional introduction of aquatic nuisance species. NANPCA further directs the U.S. Army Corps of Engineers to develop a program of research and technology for the environmentally sound control of zebra mussels in and around public facilities, and make information available on these control methods.[47]

NANPCA also reauthorized the National Sea Grant College Program Act.[48] Administered by NOAA, the Sea Grant program conducts research, outreach, and education to address marine and coastal systems, focusing among other things on aquatic invasive species particularly in the Great Lakes. The program also supports research to demonstrate ballast water technology and marine engineering advances to combat aquatic nuisance species under two efforts—Great Lakes Environmental Research Lab (GLERL) and the Cooperative Institute for Limnology and Ecosystems Research at Michigan State University (NOAA's Joint Institute partner).[49] Additional information is provided in the section on "National Oceanic and Atmospheric Administration."

National Invasive Species Act

The National Invasive Species Act of 1996 (NISA, 16 U.S.C. §§4701 *et seq.*) amended NANPCA to create a national ballast management program to prevent the introduction and spread of nonindigenous species into U.S. waters.[50] The program was modeled after the Great Lakes ballast water management program established in NANPCA.[51] NISA cites concerns about several invasive aquatic species, including the zebra mussel as well as the Eurasian ruffe (*Gymnocephalus cernuus*), mitten crab (*Eriocher sinensis*), green crab (*Carcinus maenas*), brown mussel (*Perna perna*), shellfish pathogens, and also several vegetation species, such as Eurasian watermilfoil (*Myriophylllum spicatum*), hydrilla (*Hydrilla verticillata*), anchored water hyacinth (*Eichhornia azurea*), and water hyacinth (*Eichhornia crassipes*).

In addition, NISA encouraged negotiations with foreign governments to develop and implement an international program for preventing the introduction and spread of invasive species in ballast water. NISA required the Coast Guard to report to Congress on the effectiveness of existing shoreside ballast water facilities used by crude oil tankers in the coastal trade off Alaska, as well as studies of Lake Champlain, the Chesapeake Bay, San Francisco Bay, Honolulu Harbor, the Columbia River system, and other estuaries and waters of national significance. NISA established civil and criminal penalties for certain violations, and also authorized funding for research on aquatic nuisance species prevention and control in the Chesapeake Bay, Gulf of Mexico, Pacific Coast, Atlantic Coast, and San Francisco Bay-Delta Estuary.

Alien Species Prevention and Enforcement Act

The Alien Species Prevention and Enforcement Act of 1992 (ASPEA, 39 U.S.C. §3015) defines certain categories of nonmailable plant pests and injurious animals. ASPEA does not make any new categories of plants or animals illegal to ship, but rather makes it clear that use of the U.S. mail is included among those forms of transport whose use is illegal for shipment of prohibited species. The prohibited species are those injurious animals whose movement is prohibited (under part of 18 U.S.C. §42) and those plants and animals whose shipment is prohibited (under 16 U.S.C. §3372), as well as plants covered under various plant pest and plant quarantine acts. ASPEA is administered by the U.S. Postal Service. Although ASPEA appears to do very little to prevent the introduction of invasive species, especially if the sender is unaware that the shipped items are prohibited under the above laws, it may provide for prosecutors to bring cases involving shipment of various species, including non-native invasive species, to court.

Wild Bird Conservation Act

The Wild Bird Conservation Act of 1992 (16 U.S.C. §§4901, *et seq.*) does not address introductions by non-native species, but rather conservation of birds caught in the wild in foreign countries and imported into this country. By regulating imports of certain wild birds, however, the law may have the incidental effect of reducing imports of non-native parasites and diseases that could affect wild populations of native birds.[52] It also could reduce the chance that an imported wild bird species could escape, breed, and increase to pest levels. Ten families of birds are specifically exempted from the provisions of the law, although their importation could be restricted by other applicable U.S. trade laws. FWS generally administers activities under the law.

Hawaii Tropical Forest Recovery Act

The Hawaii Tropical Forest Recovery Act of 1992 (16 U.S.C. §4503(note)) amended the International Forestry Cooperation Act (16 U.S.C. §4501(b)) to create a variety of measures to address the problems within the native forests of Hawaii. The introduction of non-native invasive species, such as pigs, goats, and mosquitoes has been a major threat to the integrity of native Hawaiian forest ecosystems. The law has several features that address these issues. USDA's Forest Service is authorized to develop a program to assist Hawaii and U.S. territories to protect native species from non-native species, and to establish biological control agents for the non- natives,[53] as well as develop plans for the Institute of Pacific Islands Forestry and for the Hawaiian tropical forests which must, among other things, provide for the study of biological control of non-native invasive species.

In addition, the law created a short-term task force of specified federal, state, and other individuals. Among its other responsibilities, the task force developed an action plan to "promote public awareness of the harm caused by introduced species" and "the benefits of fencing or other management activities for the protection of Hawaii's native plants and animals from non-native species, including the identification and priorities for the areas where these activities are appropriate."[54] Since that time, the report has been the framework for Forest Service management and research budget requests in this area.

Plant Protection Act

The Plant Protection Act of 2000 (PPA; 7 U.S.C. §§7701 *et seq.*) is the primary federal law governing plant pests in foreign and interstate commerce, covering agricultural commodities, plants, biological control organisms, articles that might be infested, means of transportation, and other pathways for moving pests. PPA consolidated several plant quarantine authorities, some dating back to the 1880s,[55] and authorizes APHIS to cooperate with states, localities and others to prevent the spread of or eradicate invasive pests and diseases. It authorizes APHIS to prohibit or restrict the importation, exportation, and the interstate movement of plants, plant products, certain biological control organisms, noxious weeds, and plant pests. It also authorizes APHIS to inspect foreign plant imports, to quarantine any state or premise infested with a new pest or noxious weed, and to cooperate with states in certain control and eradication actions. These authorities have been traditional hallmarks of U.S. plant pest regulations, and are administered by APHIS in collaboration with state departments of agriculture and their plant protection boards.

PPA gives USDA authority to use a wide range of measures to exclude alien pests or prevent the spread of new, but not widespread pests.[56] These measures include inspections, surveillance, quarantines, treatments, or destruction. USDA can develop lists of organisms that can or cannot enter the United States and goods that can be imported from specific countries, and has the authority to certify that U.S. agricultural exports meet the phytosanitary standards of other countries. USDA can require private parties to take remedial actions without cost to the government but must select the least costly, effective measure. The law also clarifies the extent of USDA's authority to regulate biological control agents and encourages the USDA, other federal agencies, and the states to facilitate biological control of pests and other invasive species, whenever feasible. The law imposes civil and criminal penalties.

PPA authorizes USDA to transfer funds from the Commodity Credit Corporation (CCC)[57] or other USDA programs to implement an emergency program to control specific plant pests of concern, subject to Office of Management of Budget (OMB) review.[58] Under some circumstances, USDA may also declare extraordinary emergencies, under which USDA can take action to control intrastate outbreaks of new pests, and has discretion to compensate growers for losses caused by the control program.[59] All states have some type of domestic quarantine laws; however, federal regulations preempt state actions in interstate commerce.[60] States also may petition the Secretary for a "special need" exception to federal rules to request permission to impose restrictions beyond what is required by APHIS.[61] In addition, PPA provides that "any person" (or state) may petition USDA to add or remove plant pests from federal regulation.[62]

Animal Health Protection Act

The Animal Health Protection Act of 2002 (AHPA, 7 U.S.C. §§8301-8302) is the primary federal law governing the protection of animal health, and gives USDA's APHIS broad authority to detect, control, or eradicate pests or diseases of livestock or poultry. APHA consolidates all of the animal quarantine and related laws, some dating back to the late 1800s, and replaces them with one statutory framework.[63] Most of the authorities contained in the consolidated AHPA were taken from existing laws, while some new provisions were added to help fully protect U.S. animal agriculture due to gaps in legal authority.

AHPA authorizes USDA to prohibit or restrict import or export of any animal or related material if necessary to prevent the spread of any livestock or poultry pest or disease, including quarantine of animals. USDA has the authority to hold, seize, treat, or destroy any animal, as well as to limit movement in interstate movement of invasive animal species. Like the Plant Protection Act, AHPA provides for emergency fund transfers and the determination of extraordinary emergencies, so that USDA can, under some circumstances, take actions within a state, and gives USDA the authority to enter into agreements with foreign governments, state governments, or other organizations. AHPA also requires compensation to farm owners based on fair market value of destroyed animals and related material. The law also authorizes USDA to transfer funds from the CCC or other USDA programs to implement an emergency control program, subject to OMB review.[64]

Noxious Weed Control and Eradication Act

The Noxious Weed Control and Eradication Act of 2004 (7 U.S.C. §§7781-7786) amended the Plant Protection Act to direct USDA to establish a grant program to provide financial and technical assistance to weed management entities to control or eradicate harmful, invasive weeds on public and private lands. The law also authorizes USDA to enter into cooperative agreements with weed management entities to fund weed eradication activities, and enable rapid response to outbreaks of noxious weeds. The law is administered by USDA's APHIS.[65]

Federal Directive

Executive Order 13112 on Invasive Species

In response to rising concern, especially in southern and western states and Hawaii, then- President Clinton signed Executive Order 13112 on Invasive Species in 1999 to prevent the introduction of invasive species, provide for their control, and minimize their impacts through better coordination of federal agency efforts.[66] Under the order, the federal government may:

> not authorize, fund, or carry out actions that it believes are likely to cause or promote the introduction or spread of invasive species in the U.S. or elsewhere unless, pursuant to guidelines that it has prescribed, the agency has determined and made public its determination that the benefits of such actions clearly outweigh the potential harm caused by invasive species; and that all feasible and prudent measures to minimize risk of harm will be taken in conjunction with the actions.

The order directs all federal agencies to address invasive species concerns, as well as to refrain from actions likely to increase invasive species problems. Interagency goals under a National Invasive Species Council (NISC) were directed to provide leadership, coordination, and oversight of federal agency activities; to encourage work with nonfederal partners; and to aid public participation. NISC, supported by an advisory committee, was directed to develop recommendations for international cooperation, promote a network to document and monitor invasive species impacts, and encourage development of an information-sharing system on invasive species. For more information, see the section on "National Invasive Species Council."

SELECTED FEDERAL AGENCIES

Numerous federal and interagency efforts share responsibilities for preventing, eradicating, and controlling invasive species.[67] Following are descriptions of how federal agencies and interagency efforts address invasive species concerns. Most of these programs also address problems of native pest species or other domestic issues. No agency devotes the majority of its resources to invasive species issues. Even so, in some cases (e.g., APHIS), efforts to address invasive species account for a substantial portion of the workload; in others (e.g., Coast Guard), invasive species are a minor share of the total program. Outside of some activities in the agricultural sector, no evidence was found of focused efforts by any agency to control the exports of U.S. species which could become harmful or invasive in the countries receiving them. A summary of the federal agencies and laws that govern invasive species is in Appendix A, based in part on information compiled by NISC.[68]

Table 2 provides a summary of federal funding for invasive species activities for FY2012, as reported by NISC. These data are self-reported by the federal agencies engaged in invasive species activities, and are not independently compiled. Further, the criteria used by agencies to compile the data are unclear and may be inconsistent across different agencies, particularly given the difficulty in separating spending related to invasive species concerns from other activities within an agency.

Overall, total spending on activities across a range of federal agencies was $2.2 billion in FY2012. Activities at the U.S. Department of Agriculture (USDA) account for the bulk of available federal funding, with nearly $1.3 billion (58% of total available funds). Activities at the Department of Homeland Security (DHS), composed of mostly border protection and security activities, accounted for about $0.7 billion (31% of total funding).

Activities across a range of agencies at the Department of Interior, including the U.S. Fish and Wildlife Service (FWS) and the National Park Service (NPS), among others, accounted for $86 million (4%). Efforts at National Oceanic and Atmospheric Administration (NOAA) under the Department of Commerce accounted for $7 million (<1% of total funding).

Table 2. Funding for Invasive Species Activities (FY2012 enacted, $ in thousands)

	DHS	DOC (NOAA)	DOD (USACE)	DOI	DOS	DOT	USAID	USDA	Total
Prevention	679,336	0	16,749	11,255	2,214	261	2,724	149,523	**862,062 (39%)**
Early Detection and Rapid Response	0	18	7,855	8,618	2,831	0	1,677	253,302	**274,301 (12%)**
Control and Management	2,100	763	87,101	40,189	10,380	0	0	479,426	**619,959 (28%)**
Research	0	2,449	3,676	9,485	2,564	0	336	228,383	**246,893 (11%)**
Restoration	0	3,126	15,364	14,442	0	0	0	37,900	**70,832 (3%)**
Education and Public Awareness	0	98	2,880	965	317	0	0	89,223	**93,483 (4%)**
Leadership / International Cooperation	0	278	1,261	733	66	118	711	50,066	**53,233 (2%)**
Total	681,436	6,732	134,886	85,687	18,372	379	5,448	1,287,823	**2,220,763**
% of Federal Spending	**30.7%**	**0.3%**	**6.1%**	**3.9%**	**0.8%**	**0.0%**	**0.2%**	**58.0%**	**100.0%**

Source: NISC, "Invasive Species Interagency Crosscut Budget," February 1, 2013, available at http://www.invasivespecies.gov/global/org_collab_budget/org_collab_budget_documents/NISC_2012_Crosscut_Budget_Summary.pdf.

Notes: Abbreviations are as follows: DHS (U.S. Department of Homeland Security); DOC (U.S. Department of Commerce); NOAA (National Oceanic and Atmospheric Administration); DOD (U.S. Department of Defense); USACE (U.S. Army Corps of Engineers); DOI (U.S. Department of the Interior); DOS (U.S. Department of State); DOT (U.S. Department of Transportation); USAID (U.S. Agency for International Development); USDA (U.S. Department of Agriculture).

The U.S. Army Corps of Engineers at the Department of Defense (DOD) accounted for about $0.1 million (6%). Activities at the U.S. State Department and the U.S. Agency for International Development (USAID) accounted for a combined total of $24 million (1%).

The budget data in Table 2 are compiled by NISC as part of the Invasive Species Interagency Performance-Based Budget (Crosscut) called for in the first National Invasive Species Management Plan. NISC developed the first Invasive Species Crosscut Budget for FY2004, followed by FY2005, FY2006, and FY2007. A modified crosscut budget was compiled for FY2010 through FY2012 following a few years hiatus.[69]

Interagency Efforts

In an effort to improve coordination across federal agencies and to promote interagency cooperation on invasive species activities, various committees have been established over the years. Some are broadly tasked with addressing a range of invasive species concerns; others are more narrowly focused on specific types of ecosystems, whether aquatic or terrestrial systems, or on specific species groups such as animals or vegetation and weeds. Although the stated missions of the various interagency committees may complement each other by specializing in a particular area, this could result in duplicative efforts and redundancies across some committee jurisdictions. Also, unlike many of the federal agencies that are engaged in invasive species activities under an explicit legal or statutory authority and, in some cases, authorized funding to implement certain programs, the interagency committees established to coordinate activities across the federal agencies often lack the authority and available funding to be able to implement certain policies and actions.

National Invasive Species Council

The National Invasive Species Council (NISC) provides high-level interdepartmental coordination of federal invasive species actions and works with other federal and non-federal groups to address invasive species issues at the national level. NISC was created by Executive Order 13112 in 1999. NISC is co-chaired by the Secretaries of the Interior, Agriculture, and Commerce. The membership of the council also consists of the Secretaries of Defense, Health and Human Services, State, Transportation, and Treasury as well as the

Administrators of the Agency for International Development (USAID) and the Environmental Protection Agency (EPA).[70]

NISC and its member agencies, supported by its advisory committee (Invasive Species Advisory Committee, ISAC),[71] are tasked with developing recommendations for international cooperation, promoting a network to document and monitor invasive species impacts, and encouraging development of an information-sharing system on invasive species. Among NISC's other duties and activities are to: [72]

- prepare, revise, and issue a national invasive species management plan;
- draft the interdepartmental invasive species performance budget;
- oversee implementation of Executive Order 13112, and review progress under the NISC Management Plan and Executive Order 13112 (as part of a five year review);
- encourage planning and action at local, tribal, state, regional, and ecosystem based level to achieve strategic goals;
- work with the Council on Environmental Quality (CEQ) to develop guidance for federal agencies pursuant to NEPA;
- work with the Department of State to provide input for international invasive species standards and cooperation; and
- facilitate development of a coordinated network among federal agencies to document, evaluate, and monitor invasive species impacts.

In 2001, NISC released its first national invasive species management plan. The 2001 plan recommended nine goals for invasive species management and, with the help of ISAC, recommended research needs and measures to minimize the risk of species introductions. The 2001 plan constituted the first federal attempt to coordinate invasive species actions over a broad range of species and habitats; across federal, state, and local governments; and with private industry, interest groups, and private individuals.[73]

Among the major features in the plan were the three key areas of prevention, early detection and rapid response, and control and management. In terms of overall federal spending for invasive species activities across all federal agencies, these three areas accounted for nearly 80% of all annual spending in FY2012: prevention ($862 million, or 39%); control and management ($620 million, 28%); and early detection and rapid response accounted for $274 million (12%). (See Table 2.)

NISC completed a five year review of Executive Order 13112 in 2005.[74] Also in 2005, the National Invasive Species Information Center (NISIC) was established.[75] The NISIC website serves as a reference gateway to information, organizations, and services about invasive species, and is maintained by USDA's National Agricultural Library. The website posts the national invasive species management plan and provides extensive links to major data bases.

NISC released its second management plan in 2008, revising the 2001 plan. The revised plan directed federal efforts from 2008 through 2012.[76] It has not yet been updated. The 2008 plan focused on five strategic goals: prevention; early detection and rapid response; control and management; restoration; and organizational collaboration. These goals are supported through efforts such as research, data and information management, education and outreach, and international cooperation.[77]

Aquatic Nuisance Species Task Force

The Aquatic Nuisance Species Task Force (ANSTF) is an intergovernmental organization, established in 1991 to implement NANPCA.[78] The ANSTF is co-chaired by FWS and NOAA, which coordinate government efforts related to nonindigenous aquatic species in the United States with those of the private sector and other North American interests.

Overall, ANSTF consists of 25 members. Of these, 13 are federal agency representatives: EPA, Coast Guard, U.S. Army Corps of Engineers, Forest Service, National Park Service, Bureau of Reclamation, Bureau of Land Management, Maritime Administration, USDA APHIS, U.S. Geological Survey, and the Department of State (along with co-chairs FWS and NOAA). The other 12 *ex officio* members include mostly regional representatives.[79] Six regional panels for the Great Lakes, Western (19 western states and Guam), Mid-Atlantic, Gulf and South Atlantic, Mississippi River Basin, and Northeast regions serve as advocates and advisory committees to ANSTF, coordinating interagency efforts to address regional priorities.

ANSTF approves comprehensive state and interstate plans for managing nonindigenous aquatic species. As of February 2013, there were 41 approved plans (38 state and 3 interstate).[80] ANSTF manages a public awareness campaign targeted toward aquatic recreation users entitled "Stop Aquatic Hitchhikers." The campaign builds on voluntary guidelines for recreational activities to highlight measures that can be taken to minimize the spread of aquatic invasive species.[81] ANSTF also has conducted studies and reports to

Congress addressing ballast water exchange, controls on vessels, and aquatic nuisance species, among other issues.

Federal Interagency Committee for Management of Noxious and Exotic Weeds

The Federal Interagency Committee for Management of Noxious and Exotic Weeds (FICMNEW) was created in 1994 through a memorandum of understanding among key federal agencies. It is composed of representatives from 16 federal agencies with invasive plant management and regulatory responsibilities. These include the Departments of Agriculture, the Interior, Transportation, Defense, and Energy as well as EPA.[82] FICMNEW fosters cooperative work on management of noxious and exotic weeds on federal lands and provides technical assistance on private lands. The committee has released several publications and issue papers, including a weed fact book, *Invasive Plants: Changing the Landscape of America, and the document Pulling Together: National Strategy for Invasive Plant Management.*[83] FICMNEW works with NISC to implement Executive Order 13122 and to coordinate federal agency activities that prevent and control invasive plants.[84]

Invasive Terrestrial Animals and Pathogens

The Invasive Terrestrial Animals and Pathogens (ITAP) is a federal scientific and technical interagency group housed at USDA to coordinate sharing of technical information for program planning and for managing invasive species.[85] It was established in 2004 by a memorandum of understanding between USDA, the Department of the Interior and the Smithsonian Institution. Other partners include the Departments of Defense, Health and Human Services, Homeland Security, State, and Transportation, as well as EPA and National Aeronautics and Space Administration. ITAP's mission is to "support and facilitate more efficient networking and sharing of technical information for program planning and coordination" among federal agencies involved with invasive species research and management. ITAP focuses on several major taxonomic groups of invasive species, and its mission parallels and complements the missions of FICMNEW and ANSTF.

Department of Agriculture

USDA has a variety of programs affecting invasive species spanning several USDA agencies,[86] and also is a statutory co-chair of NISC. Funding information compiled by NISC indicates that USDA activities relating to invasive species accounted for the majority—about $1.3 billion (58% of total) in FY2012—of total estimated funding for invasive species activities across selected federal agencies[87] (Table 2). The majority of these funds (about 75%) cover invasive species control and management, early detection and rapid response, and research activities.

Limited available information from APHIS indicates that, among agencies, more than one-half of the available invasive species spending at USDA is administered through APHIS programs.[88] Nearly 10% of USDA's spending on invasive species is administered through its Forest Service programs.[89] A breakdown of the remaining roughly one-third of the funding distributed is not available, although the majority is likely distributed through USDA's primary research agencies.

Animal and Plant Health Inspection Service

APHIS is the primary USDA agency charged with preventing plant and animal pests and diseases, including non-native invasive species, from entering the United States:[90]

> APHIS can prohibit, inspect, treat, quarantine, or require mitigation measures prior to allowing entry of plant species, plant pests, biological control organisms, animals, animal products and by- products, or their host commodities or conveyances. APHIS is involved with overseas control and eradication of some invasive pest species. APHIS also regulates the importation/exportation of veterinary biological products intended to treat animal disease.

APHIS is responsible for protecting U.S. agriculture from domestic and foreign pests and diseases, responding to domestic animal and plant health problems, and facilitating agricultural trade. As part of APHIS' regulatory framework, the agency regulates certain animals and animal products to guard against the introduction of animal diseases into the United States,[91] and regulates certain plants and plant products prohibiting or restricting the importation of plants, plant parts, and plant products into the United States.[92] APHIS also lists noxious weeds that may be a concern involving the importation and interstate movement of plants and plant products.[93] The text box below provides a partial listing of some of the plant and wildlife programs and ongoing efforts at USDA—primarily addressing concerns to U.S. agriculture. Many of these plant and animal species are invasive.

Table 3 provides an overview of APHIS program-level spending on invasive species efforts from a special compilation conducted by USDA for FY2010.

Selected APHIS Plant and Wildlife Pest Programs/Efforts

Following is a partial listing of some ongoing USDA plant and wildlife programs, most of which address species primarily regarded as a concern to U.S. agricultural production, including fisheries and aquaculture. Many of these plant and animal species are invasive.

- **Plant Diseases**: Black stem rust/barberry; chrysanthemum white rust; citrus diseases (including citrus canker, citrus greening, and Asian citrus psyllid); European larch canker; gladiolus rust; karnal bunt; sudden oak death; plum pox; potato diseases (such as potato virus Y strains, potato wart, and ralstonia); soybean rust; and thousand cankers disease.
- **Weeds**: Anchored waterhyacinth; giant salvinia; giant hogweed; hydrilla; melaleuca (or broadleaf paper bark tree); *Mikania micrantha*; onionweed; old world climbing fern; maidenhair creeper; tropical soda apple; tropical spiderwort (also known as Benghal dayflower); and witchweed (among many other aquatic and parasitic noxious weeds).
- **Nematodes**: Golden Nematodes and pale cyst nematode.
- **Mollusks**: Giant African land snail; temperate terrestrial gastropods; and zebra and quagga mussels (latter mostly addressed by NOAA and state department of fish and game).
- **Insects and Mites**: Asian longhorned beetle; brown marmorated stink bug; elm seed bug; cotton pests (such as boll weevil and pink bollworm); spotted wing drosophila; emerald ash borer; European grapevine moth; false codling moth; fruit flies (especially within the genera Anastrepha, Bactrocera, and Ceratitis); grasshopper/ Mormon cricket; gypsy moth; imported fire ant; Japanese beetle; Khapra beetle (or cabinet beetle); light brown apple moth; panicle rice mite; pine shoot beetle; and palm weevils. (European honey bees, introduced to the United States, are also considered non-native.)
- **Other Wildlife**: Brown tree snake; European starling; feral swine; nutria; and Gambian pouch rat.

Sources: Compiled by CRS from various USDA sources: APHIS, Plant Pest Program Information (http://www.aphis.usda.gov/plant_health/plant_pest_info/); APHIS, Wildlife Services "Invasive Species" (http://www.aphis.usda.gov/wildlife_damage/invasive_species.shtml); APHIS, "Outlawed! Federal Noxious Weeds: The Aquatics" (http://www. aphis.usda.gov/plant_health/plant_pest_info/weeds/downloads/NWposter.pdf); and "Federal Noxious Weeds List" (http://www.aphis.usda. gov/plant_health/plant_pest_info/weeds/downloads/weedlist.pdf), as well as APHIS, "Hungry Pest" website; http://www.hungrypests.com. More detailed information about each of these programs as well as a full listing of weeds under the "Federal Noxious Weeds" program is at various USDA websites.

APHIS conducts program delivery, research, and other activities through its regional and state offices, the National Wildlife Research Center and its field stations, as well as through its national programs. It has a number of ongoing efforts targeting certain plant pest concerns for key invasive species that are known to harm agricultural production.[94] APHIS also administers the Plant Epidemiology and Risk Analysis Laboratory (PERAL), whose scientists and professionals conduct Plant Protection and Quarantine (PPQ) analyses for pest risks. PERAL is responsible for providing essential scientific support to risk-based policy-making across a broad range of phytosanitary issues.

APHIS's Wildlife Services activities target introduced and invasive animal species of concern, including brown tree snakes, Gambian rats, nutria, coqui frogs, pigeons, starlings, house sparrows, feral pigs, and Burmese pythons.[95] APHIS' sVeterinary Services activities include the National Animal Health Laboratory Network, which is a state-federal cooperative effort including the APHIS National Veterinary Services Laboratories, and provide reference and confirmatory laboratory services including training, proficiency testing, and prototypes for diagnostic tests. For example, more than 40 laboratories have been trained and proficiency-tested to perform foot and mouth disease, avian influenza, and exotic Newcastle surveillance diagnostics, among other animal-related diseases.

Control methods used by the agency include providing advice to individuals and to municipal, state or federal agencies on a wide variety of preventative, non-lethal control methods. Control of predatory animals, native or non-native, is largely carried out by lethal means, including hunting, trapping, and poisoning. The agency publishes annual Program Data Reports to inform the public about its wildlife damage management activities.[96]

Table 3. Estimated Spending on APHIS Invasive Species Programs, FY2010

Line Item	FY2010 ($1,000)
Agricultural Quarantine Inspection (Approp)	29,000
Cattle Ticks	13,157
Foreign animal diseases/Foot-and-mouth disease FAD/FMD	4,004
Fruit Fly Exclusion and Detection	62,920
Import/Export	13,298
Overseas Technical and Trade Operations	16,172
Screwworm	27,714
Tropical Bont Tick	429
Animal Health Monitoring & Surveillance	121,667
Animal & Plant Health Reg. Enforcement	13,983
Avian Influenza	60,243
Emergency Management Systems	15,794
National Veterinary Stockpile	3,757
Pest Detection	28,113
Select Agents	5,176
Aquaculture	6,560
Biological Control.	10,467
Brucellosis	9,707
Chronic Wasting Disease	16,875
Contingency Funds	2,058
Cotton Pest	23,390
Emerging Plant Pests	158,769
Golden Nematode	831
Gypsy Moth	5,420
Imported Fire Ant	1,902
Johne's Disease	6,876
Noxious Weeds	1,990
Plum Pox	2,206
Pseudorabies	2,510
Scrapie	17,906
Tuberculosis	16,764
Rabies	23,810
Brown Tree Snake	1.3
Witchweed	1,517
Biotechnology Regulatory Service	13,050
Total	**$738,036**

Source: USDA, APHIS (data request by CRS staff, May 2, 2011). Estimated APHIS program-level spending on invasive species efforts from a special compilation conducted by USDA for FY2010.

The agency has memoranda of understanding and other cooperative agreements with FWS, the National Park Service, the Bureau of Land Management, the Forest Service, and state natural resource agencies to help protect natural resources, including wildlife and threatened or endangered species, from loss of life, habitat, or food supply due to the activities of other species. The agency also addresses damage problems caused by such non-native species as nutria (*Myocastor coypus*), European starlings (*Sturnus vulgaris*), and monk parakeets (*Myiopsitta monachus*) and is also charged with monitoring and controlling the brown tree snake.

The two primary laws administered by APHIS are the Plant Health Protection Act (7 U.S.C. §§7701-7721) and the Animal Health Protection Act (7 U.S.C. §§8301-8322). Other laws and statutes governing APHIS activities relating to invasive species also include the Agricultural Bioterrorism Act (7 U.S.C. §8401); the Animal Damage Control Act (7 U.S.C. §§426 *et seq.*); the Federal Seed Act (7 U.S.C. §§1551 *et seq.*); the Federal Noxious Weed Act (7 U.S.C. §2814); and the Noxious Weed Control and Eradication Act of 2004 (7 U.S.C. §§7781-7786).

Farm Service Agency

The Farm Service Agency (FSA) administers USDA's Conservation Reserve Program (CRP).[97] CRP is a voluntary program that helps agricultural producers and landowners use environmentally sensitive lands (such as lands that are highly erodible) for conservation benefits, including weed control (including noxious weeds) and invasive species, insects, pests, and other undesirable species on enrolled lands.[98] Accordingly, the primary statute governing FSA relating to invasive species are provisions regarding CRP (16 U.S.C. §§3838a, 3832).

Foreign Agricultural Service

The Foreign Agricultural Service (FAS), working with APHIS, helps provide invasive species technical assistance to foreign countries seeking to import raw agricultural commodities and food products into the United States to ensure they do not also inadvertently introduce plant and animal pest and disease concerns, including invasive species.

Forest Service

The Forest Service manages invasive activities on 193 million acres of National Forests and grasslands. As part of its forest and resource management activities, it has numerous programs intended to prevent invasive species introduction and spread, controlling the most threatening invasive species, monitoring to detect newly introduced species, and restoring ecosystems damaged by invasive plants, insects or pathogens. The agency's activities in the National Forest System (NFS) are intended to improve forest management by preventing, controlling, and eradicating aquatic and terrestrial invasive species (including invasive plants, pathogens, vertebrates, and invertebrates), as well as monitoring to detect newly introduced species, and restoring ecosystems damaged by invasive plants, insects or pathogens.[99] Some Forest Service activities regarding invasive species are:[100]

- respond to nationwide threats to forest ecosystems from non-native invasive species: insects, pathogens and plants;
- support the establishment of Cooperative Weed Management Areas and also Cooperative Invasive Species Management Areas;
- develop a Forest Service Manual (FSM 2900) for invasive species management on the NFS, and also a NFS Invasive Species Management Handbook;
- establish an Early Detection and Rapid Response (EDRR) initiative and also an Invasive Insects Early Detection Program;
- conduct training, and provide funding and technology for invasive species work;
- develop policies regarding both native and invasive species management in national forests; and
- conduct data management, recordkeeping and reporting, and research.

To support these efforts, the FS conducts research focused on invasive plant species, including ecological studies to support restoration of sites after treatment of exotic weeds, as well as control Miconia sp. and other invasive plants in Hawaii; kudzu (*Pueraria montana var. lobata*) in the southern United States; yellow starthistle (*Centaurea solstitialis*), spotted knapweed (*Centaurea maculosa*), and leafy spurge (*Euphorbia esula*) in Idaho; among other non-native invasive species. In addition, the Forest Service seeks to control and mitigate the impacts from harmful non-native invasive insects, such as the Asian longhorned beetle (*Anoplophera glabripennis*), gypsy moth (*Lymantria dispar*), hemlock woolly adelgid (*Adelges tsugae*), and browntail moth (*Euproctis chrysorrhoea*). The agency conducts research on such tree diseases as butternut canker and sudden oak death syndrome, and works to find and develop trees genetically resistant to Dutch elm disease, pitch canker, chestnut blight, and white pine blister rust.

The Forest Service works closely with state agencies, private landowners, and tribal governments on prevention and control activities, and provides funding and technical assistance through its state and private forestry programs.

Primary laws governing Forest Service programs relating to invasive species include[101] the Organic Administration Act (16 U.S.C. §551), Multiple-Use Sustained-Yield Act (16 U.S.C. §§528-531), Forest and Rangeland Renewable Resources Planning Act (16 U.S.C. §§1671 et seq.), Federal Noxious Weed Act (7 U.S.C. §2814), Public Rangelands Improvement Act (43 U.S.C. §§1901-1908), Federal Land Policy and Management Act (43 U.S.C. §1701), Cooperative Forestry Assistance Act (16 U.S.C. §§2101-2111), among others. Many of these authorities do not focus strictly on invasive species management, but also apply to other Forest Service activities including rangeland management, research, or public use activities.

Research, Education, and Economics Agencies

USDA's Research, Education, and Economics agencies are responsible for research, analysis, and data collection within USDA. Three agencies support research on invasive species: Agricultural Research Service (ARS), Economic Research Service (ERS), and National Institute of Food and Agriculture (NIFA).[102]

Agricultural Research Service

ARS is USDA's chief in-house research agency, and provides scientific and technical support for its regulatory agencies, including APHIS. ARS has significant involvement in pest management, including invasive species, through its research infrastructure. Its personnel and facilities in domestic and foreign laboratories provide support to other agencies, organizations, and state governments. Some of the agency's activities regarding invasive species are:[103]

- maintain overseas biological control laboratories and quarantine facilities, and develop procedures relating to conducting tests for the release of biological control agents in accordance with the requirements of NEPA;[104]
- maintain and manage e-Government and public communication initiatives, such as the website, www.invasivespeciesinfo.gov, through the agency's National Invasive Species Information Center at the

National Agricultural Library within ARS, among other public outreach and educational activities;

- conduct biological control host-specificity testing to support research efforts in Integrated Pest Management (IPM);[105]
- conduct area-wide pest management programs, including demonstrations of IPM and other systems that employ biologically-based or pest-specific methods to reduce use of certain chemical pesticides;
- develop technologies and approaches to reduce non-target effects associated with conventional pesticides; and
- provide support for the NISC and ITAP.

The wide-ranging research activities of ARS include the detection, identification, characterization, prevention, monitoring, and control of plant diseases, insects, weeds, and animal diseases.

Economic Research Service

ERS, the USDA's in-house economic research agency, contributes to USDA's invasive species efforts through its Program of Research on the Economics of Invasive Species Management (PREISM), as well as its programs on pesticide use and pest management economic research and analysis program. PREISM was initiated in 2003 to examine the economics of managing invasive pests in increasingly global agricultural markets. Through PREISM, ERS has funded a competitive awards program that focuses on national decision-making concerning invasive species of agricultural significance or affecting, or affected by, USDA programs. For example, between FY2003-FY2008, ERS disbursed $6.8 million for research on the economics of invasive species.[106] More recent information indicates that no funds have been allocated in subsequent years. Information on PREISM and outputs of PREISM-funded projects are reported at USDA's website.[107]

In addition, ERS research on pesticide use provides information used to administer the integrated pest management program, Food Quality Protection Act (FQPA)[108] implementation, and also invasive species programs.

National Institute of Food and Agriculture

NIFA distributes federal funds to support research and extension programs at the land grant colleges of agriculture in every state.[109] NIFA supports research, education, and extension programs in the land grant university system and other partner organizations; it does not perform actual research, education, or extension but instead helps fund programs at the state and local level. NIFA allocates some funds to each state according to formulas spelled out in authorizing laws, and distributes the rest through various competitive grant programs. Information on USDA programs that could be used to fund and support invasive species related projects is available at USDA's website.[110]

NIFA maintains two national plant and animal diagnostic laboratory networks to detect and report pathogens of national interest, and to provide timely information and training to state university diagnostic laboratories. These include the National Plant Diagnostic Network[111] and the National Animal Health Laboratory Network.[112] NIFA also supports ongoing IPM work, pesticide use and pest management, training, and other extension and education programs. NIFA supports activities in the Technical Advisory Group for the Biological Control of Weeds.[113] NIFA also supports state- level research on invasive species, and extension programs to help farmers, ranchers, and private landowners to adopt cost-effective, environmentally safe controls for invasive species.

Natural Resources Conservation Service

The Natural Resources Conservation Service (NRCS) provides technical assistance to cooperating landowners and federal agencies (such as the Forest Service and Bureau of Land Management) on adopting conservation practices on agricultural land, including rangeland. The agency also promotes conservation planning through many of its farmland conservation programs that provide both technical and financial assistance to farmers and landowners. Some NRCS activities regarding invasive species are as follows.[114]

- Provide U.S. private landowners with financial and technical assistance to control and/or eradicate invasive plants in an effort to maintain the desired vegetation (e.g., food crops and forage), to maintain the desired characteristics of the land (e.g., wetland open water), and to diminish invasive plants spreading to neighboring lands.[115]
- Develop NRCS Conservation Practice Standards, consisting of 170 practice standards to guide applying conservation technology on the land and minimum levels for application of the technology.
- Operate NRCS Plant Materials Centers nationwide to cultivate and provide seed stock of plants that are used for vegetative conservation practices within various geographical regions. These centers

encourage use of native plants for restoration, reclamation, and conservation practice uses, and support studies nationwide to control or suppress weeds and find suitable replacements for invasive species once control is achieved.[116] In addition to the National Plant Materials Center (located in Beltsville, Maryland), other materials centers serve several major land resource areas.[117]

- Support the PLANTS Database (http://plants.usda.gov) to determine beneficial plants that do well within a particular geographical location, as well as to list plants that should not be planted within a particular environment (e.g., federal and state noxious weed lists).[118]

More information is in the agency's "NRCS Invasive Species Policy."[119]

Among the laws that govern NRCS activities relating to invasive species are the Soil Conservation and Domestic Allotment Act (16 U.S.C. §590(a)-590(f)) and various farmland conservation provisions in various omnibus farm bill laws.

Department of Commerce

National Oceanic and Atmospheric Administration

The National Oceanic and Atmospheric Administration (NOAA) is a statutory co-chair of both the interagency NISC and ANSTF, and administers a variety of programs aimed at expanding and coordinating prevention, early detection, rapid response, control, and monitoring programs nationwide. NOAA is responsible for supporting research and monitoring efforts on the effects of aquatic invasive species on ecosystems and socioeconomic factors. It also assists regions and states by providing technical support and best management practices to prevent and contain invasive species. NOAA sub-agencies, including the National Ocean Service and the National Marine Fisheries Service, are involved in both prevention and control activities. In addition, the National Ocean Service monitors coastal areas for the presence of nonindigenous species.[120]

Funding for NOAA's invasive species activities totaled $6.7 million in FY2012 (less than one half of one percent of total federal funding for invasive species), and was used mostly for research and restoration activities. (See Table 2.)

NOAA's Sea Grant programs on invasive species focus on marine systems and the Great Lakes, through funding of research, education, and outreach to address threats from invasive species.[121] Through this program, NOAA has supported research on ballast water technology and marine engineering advances to combat aquatic nuisance species under two efforts—NOAA's Great Lakes Environmental Research Lab (GLERL) and at the Cooperative Institute for Limnology and Ecosystems Research at Michigan State University (NOAA's Joint Institute partner). Regarding invasive species, GLERL targets both the prevention and control to stop the inflow and spread of new aquatic organisms, with particular emphasis on ship ballast, and also understanding and minimizing the ecological and economic impacts of recent species invasions, especially the on- going secondary effects of zebra mussels.[122] GLERL also leads investigations of invasive species impacts on the Great Lakes ecosystem, focusing on zebra mussels and other recent invaders. Other program efforts support research on the biology of non-native invasive species; impacts of invasive species on ecosystems, including socioeconomic analysis of costs and benefits; control and mitigation options; prevention of new introductions; and reduction in the spread of established populations of harmful non-native species. The program also funded a Nationwide Zebra Mussel Training Initiative to provide technical services outside the coastal and Great Lakes areas and help provide inland states with a knowledge base for creating state and regional programs.

The primary law governing NOAA's role in addressing invasive species is the Nonindigenous Aquatic Nuisance Prevention and Control Act (16 U.S.C. §4701, *et seq.*), as amended.

Department of Defense

The Department of Defense (DOD) manages over 25 million acres of lands within military installations and engages in management and control of invasive non-native species. DOD controls and manages invasive species in accordance with individual plans governing each installation or base. The goals of DOD's Invasive Species Management Program are prevention, control of invasive species on military installations, and restoration using native plants. The Armed Forces Pest Management Board coordinates DOD activities to prevent and control the spread of invasive species, including the brown tree snake and noxious weeds, on, to, or from military bases. DOD is also promulgating joint regulations with EPA to set national performance standards that will require the use of marine pollution control devices to control discharges incidental to the normal operation of armed forces vessels.[123]

Funding for DOD invasive species efforts totaled $135 million in FY2012 (about 6% of all federal funding for invasive species), most of which was used for control and management activities through the U.S. Army Corps of Engineers (Table 2.)

Army Corps of Engineers[124]

The U.S. Army Corps of Engineers (USACE) supports a range of invasive species efforts. The Corps generally undertakes efforts to prevent or reduce the establishment of invasive species at its projects pursuant to its national USACE Invasive Species Policy.[125] Invasive species work within individual projects is typically funded through Operations and Maintenance (O&M) funding for each project; the nature of the work at the project level is addressed in project planning documents. Other USACE activities (e.g., regulatory activities) must also take invasive species into consideration.

USACE also has specific programs that address subcategories of invasive species. The Aquatic Plant Control Program provides cost-shared assistance to states for aquatic plant management that is not a part of routine federal project maintenance.[126] At full federal costs, USACE administers an Aquatic Nuisance Species Research Program, which develops methods and provides general guidance and research assistance on invasive species control strategies.[127] Finally, USACE fully funds control of aquatic plants, predominantly invasive species, in waterways in certain southeastern states through its Removal of Aquatic Growth Program.[128]

In some cases, Congress has authorized Corps invasive species control efforts at specific federal water resource projects administered by USACE in the form of physical construction and other project-level authorities. The most notable example is control of Asian river carp (*Hypophthalmichthys sp.*) in the Chicago area.[129] In that case Congress authorized and funded USACE's construction and operation of underwater electric barriers to help prevent the encroachment of Asian river carp into the Great Lakes.[130] Congress also authorized USACE to study and carry out other means to control Asian river carp in the same area, as well as a larger study of potential methods to prevent the transfer of aquatic nuisance species between the Great Lakes and Mississippi River basin.[131]

Department of Health and Human Services

Although not the focus of this report, certain invasive species can harm both ecosystems and human health. Some invasive pathogens and parasites infect humans exclusively.

The Centers for Disease Control and Prevention (CDC) within the Department of Health and Human Services (HHS) addresses zoonotic diseases,[132] emerging diseases and those with unknown causes.[133] The National Institute of Health (NIH) provides research support for zoonotic diseases and bioterrorism preparedness. In addition, NISC coordinates actions on animal diseases and human diseases, including diseases that are transmitted between human and animals. The work of various NISC agencies (such as NPS, NOAA, USDA, EPA, and DoD) includes zoonotic disease surveillance and prevention.[134]

Department of Homeland Security

After USDA, the Department of Homeland Security (DHS) accounts for much of the aggregate estimated federal funding for invasive species activities, mostly through prevention as part of its border protection activities at various sub-agencies. In FY2012, DHS funding for invasive species efforts totaled $681 million (about 31% of total federal funding for invasive species). (**Table 2**.)

Coast Guard

The Coast Guard, an agency of DHS, works with other federal agencies to develop and implement a national ballast water management program aimed at preventing the unintentional introduction and dispersal of nonindigenous aquatic species into waters of the United States from ship ballast water. As mandated by NISA (16 U.S.C. §§4701 *et seq*., amending NANPCA), this national program became mandatory in 1999.[135] Under the program, the Coast Guard has established quantitative performance standards for ballast water treatment; protocols for testing, verifying, and reporting on treatment technologies; and a program to facilitate experimental shipboard installation and operation of promising technologies.[136] The final rule to implement NISA/NANCPA was promulgated in March 2012 and requires the Coast Guard to ensure to the maximum extent practicable that aquatic nuisance species are not discharged into waters of the United States from vessels. It also requires vessels to carry out management practices necessary to reduce the probability of unintentional discharges of invasive species resulting from ship operations other than ballast water discharge, and allows for the approval of certain alternative ballast water management methods, in some cases.[137]

Customs and Border Protection

Customs and Border Protection (CBP) is responsible for securing the border and facilitating lawful international trade and travel while enforcing hundreds of U.S. laws and regulations. CBP guards nearly 7,000 miles of land border the United States shares with Canada and Mexico and 2,000 miles of coastal waters surrounding the Florida peninsula and off the coast of Southern California. The agency also protects 95,000 miles of maritime border in partnership with the U.S. Coast Guard.[138] Among its many border protection responsibilities, CBP works with USDA and the Department of Interior to enforce laws prohibiting or limiting the entry of invasive species. CBP also works with wildlife inspectors at FWS, USDA, and other federal trade inspection agencies to facilitate the detection and disruption of wildlife trafficking.[139] CBP agriculture specialists prevent the entry of harmful plant pests and exotic foreign animal diseases and confront emerging threats in agro- and bioterrorism. As part of its role in enforcing plant and animal regulations, CBP will detain, where necessary, imported or exported products pending their clearance by agency inspectors. CBP also supports the removal of invasive plants that interfere with border area surveillance.[140] CBP states they regularly discover pests at U.S. ports of entry, many that are potentially harmful to agricultural and natural resources; they also hold materials—plant, meat, animal byproduct, and soil—for quarantine.[141]

Federal Emergency Management Agency

As part of the Federal Emergency Management Agency's (FEMA) "Emergency Support Function," the agency works with other federal agencies and also supports state, tribal, and local authorities to, among other things, "control and eradicate outbreaks of animal/zoonotic disease, exotic plant pests, or invasive plant pest infestations."[142]

Immigration and Customs Enforcement

Immigration and Customs Enforcement (ICE) is DHS's principal investigative arm and was created in 2003 through a merger of the investigative and interior enforcement elements of the former U.S. Customs Service and the Immigration and Naturalization Service. ICE's primary mission is to "promote homeland security and public safety through the criminal and civil enforcement of federal laws governing border control, customs, trade, and immigration."[143] As part of its overall border protection responsibilities, ICE works with other federal agencies such as FWS, NOAA, and CBP to enforce U.S. trade laws, including laws that prohibit illegal wildlife trafficking.[144]

Department of the Interior

The Department of the Interior (DOI) is the largest land and water manager in the United States, and invasive species are an aspect of most of its programmatic responsibilities and missions. DOI is also a statutory co-chair to both NISC and ANSTF. The federal staff of NISC and the non- federal staff of ISAC are both housed and administered within DOI.

Even so, compared to some other federal agencies, DOI agencies accounts for a relatively small share of aggregate estimated federal funding for invasive species activities, reported at $85.7 million (about 4% of total federal funding for invasive species) in FY2012. DOI invasive species activities focus on control and management, restoration, and prevention. (See Table 2.)

Bureau of Indian Affairs

The Bureau of Indian Affairs (BIA) is responsible for protecting and improving the trust assets of Indian tribes while maintaining a relationship within the spirit of self-governance. The BIA, through exotic weed eradication and other programs, helps support the management of non-native invasive species on Indian lands (e.g., reservations, pueblos, rancherias, communities). In addition, BIA's Noxious Weed Control program supports resource protection on trust lands in compliance with various laws[145] and also provides education, direction and technical guidance to individual Indians, non-Indian farmers and ranchers, Indian tribes and Alaska natives involved in controlling noxious weeds.

BIA also cooperates in DOI's Invasive Species Crosscut Initiative and participates in three of the Area Invasive Plant Initiatives: Rio Grande (tamarisk or salt cedar (*Tamarix sp.*)); Northern Great Plains (leafy spurge, yellow star thistle); and Florida (melaleuca, tropical soda apple (*Solanum viarum*), and Brazilian pepper (*Schinus terebinthifolius*)). The BIA funds tribal projects in all three of these initiatives.

Bureau of Land Management

The Bureau of Land Management (BLM) focuses its invasive species efforts primarily on controlling non-native plants on the 250 million acres it manages, primarily in western states and Alaska.[146] BLM works with state, federal and local partners to reduce the spread of invasive species, and focuses on early detection of and rapid response to new invasions and to reduce the need for larger, more expensive treatments. BLM's action plan details its strategy to prevent and control the spread of noxious weeds on public lands; the goals of the plan roughly parallel those of NISC.

BLM also provides for education and cooperative efforts with various states to control exotic weeds.[147] It maintains a cooperative research relationship with most USDA agencies (APHIS, Forest Service, and USDA's research agencies). BLM also is responsible for protecting, controlling, and managing populations of wild horses and burros which, although not native, have a legally protected status.[148] APHIS, through its Wildlife Services program, helps to regulate animal pests (primarily predator control), whether native (such as coyotes (*Canis latrans*)) or non-native (such as Eurasian collared doves (*Streptopelia decaocto*)), on BLM land. On its grazing lands, BLM requires that non-native plant species be used only when native species are not available in sufficient quantities or are incapable of maintaining or achieving properly functioning conditions and biological health.

Bureau of Reclamation[149]

The Bureau of Reclamation's research, prevention, detection, and control programs address primarily the pests of aquatic systems such as reservoirs, canals, pipelines, and rivers. Such pests include both plants and animals, including tamarisk (salt-cedar), hydrilla, Eurasian watermilfoil, giant salvinia (*Salvinia molesta*), quagga and zebra mussels, and Asian river clams (*Corbicula fluminea*). Their presence results in loss of irrigation water, blocked waterways, impediments to navigation, and lost recreational opportunities. To control these pests, the agency uses biological control agents and pesticide application. Ongoing projects include research and development for eradicating or controlling invasive mussels, insect control for weed species; grass carp (itself a non-native species) for control of certain aquatic weeds, and use of herbicide meters to match herbicide flow to fluctuating water levels. The agency also maps the movement of certain invasive species (e.g., giant salvinia, hydrilla, and quagga and zebra mussels), and works with partners in Cooperative Weed Management Areas in western states to identify and control weeds. It works to improve control methods and basic knowledge of non-native invasive species, and to develop methods to restore areas with tamarisk infestation. Mitigating the effects of zebra and quagga mussels is among its top research and development priorities.[150] It also works with federal agencies, state and local governments, and others including Mexican officials on cross-border weed infestations. It also supports the work of NISC.

Fish and Wildlife Service

The U.S. Fish and Wildlife Service (FWS) is authorized to help prevent the introduction and spread of invasive species in general and, on its own lands, to control established non-native invasive species.[151] FWS maintains numerous programs covering fisheries, endangered species, habitat conservation, refuge operations and maintenance, and international affairs. FWS also enforces laws and regulations concerning the importation of injurious wildlife species.

FWS wildlife inspectors also help prevent the entry of invasive species, working with other federal agencies (USDA, NOAA, and CBP) and state and local governments. Several other programs providing grants to states, territories, and tribes may be used for prevention, control, or eradication of invasive species. Annual spending on invasive species under these programs will vary depending on the applications received. FWS allocated \$219.1 million nationally for wildlife and habitat management in FY2012, within the 150-million-acre National Wildlife Refuge System (NWRS). Many refuges spend substantial portions of their budgets on the control of such non-native invasive species such as feral pigs (*Sus scrofa domesticus*), Burmese pythons (*Python bivittatus*), melaleuca, tamarisk (or salt cedar), and purple loosestrife. The agency attempts to minimize the use of pesticides and herbicides in these efforts. FWS also has five Invasive Species Strike Teams available to respond rapidly to new infestations of invasive species before they become established.

Under wildlife protection statutes, such as ESA, the agency's authority to protect domestic ecosystems is indirect or general, meaning the agency sometimes finds itself at odds with other interests, particularly those wishing to introduce various species for sport fishing or hunting. Its broad authority under ESA allows the agency to act if a proposed introduction or other activity seems likely to harm a protected species.[152] Its spending on harmful non-native species occurs in a number of its programs, including fisheries (e.g., Aquatic Nuisance Species and brown tree snake programs), endangered species, Partners for Fish and Wildlife (e.g., some funding earmarked by Congress for special projects), habitat conservation (coastal program), refuge operations and maintenance (e.g., inventory, mapping, and monitoring of invasive species), and also international affairs.

The primary laws governing FWS's role in addressing invasive species are: Lacey Act (18 U.S.C. §§42-43; 16 U.S.C. §§3371-3378); Endangered Species Act (16 U.S.C. §1531-1543); Nonindigenous Aquatic Nuisance Prevention and Control Act (16 U.S.C. §§4701, *et seq.*); and Wild Bird Conservation Act (16 U.S.C. §§4901, *et seq.*); among other authorities.

Geological Survey

The U.S. Geological Survey (USGS) conducts and supports research that assists resource managers in the control of invasive species and restoration of affected areas. Its invasive species program focuses on "early detection and assessment of newly established invaders, monitoring of invading populations; improving understanding of the ecology of invaders and factors in the resistance of habitats to invasion; and development and testing of prevention, management, and control methods."[153] USGS has focused its research in recent years on a number of highly invasive species in the Great Lakes and eastern waterways and wetlands; in riparian ecosystems; invasive species in Hawaii and Florida; large constrictor snakes; and invasive plants on western rangelands. USGS also manages the national Nonindigenous Aquatic Species Database, as well as several regional databases (e.g., Hawaii, Colorado plateau, and northern prairie) and manages a nonindigenous aquatic species website.

USGS had also begun efforts to compile a central database, called the National Biological Information Infrastructure (NBII) to identify, document, disseminate, and integrate information about the nation's biological resources generally, including its invasive species.[154] The NBII's purpose was to facilitate access to data and information on U.S. biological resources, as well as early detection and predictive modeling efforts for invasive species in each refuge. However, due to FY2012 budgetary concerns, the program was terminated in January 2012, along with many other federal programs identified for termination or reduction.

National Park Service[155]

Over 6,500 non-native invasive species have been found on National Park Service (NPS) lands and waters. Of these, about 650 are in marine environments.[156] NPS uses an integrated pest management approach to control exotic species,[157] and has a number of programs targeted to specific sites or species. For example, in Hawaii, NPS has designated Special Ecological Areas that best represent native Hawaiian systems, and these areas are managed for the removal of exotic species. Elsewhere, NPS's Lionfish Response Plan is a species-specific program that addresses problems with invasive lionfish (*Pterois volitans*) at several units on the Atlantic seaboard and in the Gulf of Mexico, through public education, monitoring, and removal.[158]

The agency targets quagga and zebra mussels in at least 54 park units, taking actions such as inspecting and cleaning boats at ramps and mooring locations. Some units have special regulations to minimize the potential for spreading zebra mussels and other aquatic nuisance species. For example, regulations for the St. Croix (WI) National Scenic Riverway address zebra mussels, purple loosestrife, and Eurasian watermilfoil as aquatic nuisance species.[159] In the western United States, some of the non-native invasive species of concern are leafy spurge, knapweed (*Centaurea sp.*), Japanese brome (*Bromus japonica*), and cheatgrass (*B. tectorum*).

NPS is authorized to regulate fishing on its lands and prohibits the possession or use of live or dead minnows or other bait fish, amphibians, non-preserved fish eggs or fish roe as bait for fishing, except in designated waters.[160] Waters which may be so designated are limited to those where invasive species are already established, where scientific data indicate that the introduction of additional numbers or types of non-native species would not hurt populations of native species, and where park management plans do not call for elimination of non-native species.

NPS uses Exotic Plant Management Teams (EPMTs) for rapid response to invasive plants on units of the National Park System. The teams are explicitly modeled on teams used to fight fires. The team approach provides quick response and consistent application of techniques. The EPMTs provide a personnel resource not otherwise available to these parks, and reduces the need for individual parks to procure and maintain expensive equipment.[161] In an FY2011 report, the EPMTs reported that they inventoried 2,164,232 acres for exotic plants, uncovering 9,589 acres of infestation, of which 8,453 acres were treated (or re-treated) for exotic plant infestations.[162] The EMPTs have claimed elimination of two invasive plant species at Haleakala National Park (HI) and of all exotic plants at Loggerhead Key at Dry Tortugas National Park (FL), among other accomplishments. Lake Mead National Recreation Area is the focal point for an EMPT whose tasks include controlling salt cedar. EPMTs train personnel from other federal agencies with these methods. Their list of partners has included such disparate entities as National Wildlife Refuges, the Navajo Nation, Florida Power and Light, state departments of transportation, and the Nature Conservancy.

The primary laws governing NPS's role in addressing invasive species are the National Park System Organic Act (16 U.S.C. §§1 *et seq.*; 16 U.S.C. §594), Endangered Species Act (16 U.S.C. §1531 *et seq.*), Noxious Weed Control and Eradication Act (7 U.S.C. §§7781-7786), Plant Protection Act (7 U.S.C. §7701 *et seq.*), National Invasive Species Act (16 U.S.C. 4701), Nonindigenous Aquatic Nuisance Prevention and Control Act (16 U.S.C. §§4701), and Animal Damage Control Act (7 U.S.C. §§426-426c), among other authorities.

Office of Surface Mining Reclamation and Enforcement[163]

The Office of Surface Mining Reclamation and Enforcement (OSM) is responsible for regulating coal mining in the United States under the Surface Mining Control and Reclamation Act (SMCRA).[164] OSM regulations govern the permitting of coal mining operations, including the reclamation of lands disturbed by surface coal mining and surface impacts incident to underground coal mining. As part of this responsibility, OSM addresses invasive species in mine reclamation areas. These regulations allow the use of introduced species in the revegetation of disturbed or impacted lands, within certain limitations. The substitution of introduced species for native species is subject to approval by OSM, or by the state if it has regulatory primacy under SMCRA.

To reclaim disturbed or impacted surface lands affected by coal mining, OSM regulations require permittees subject to SMCRA to establish a diverse, effective, and permanent vegetative cover comprised of native species, or introduced species where desirable and necessary to achieve the approved post-mining land use. Introduced species also may be used to provide a quick-growing, temporary, stabilizing cover while measures utilizing native species are established. Furthermore, introduced species used in either instance must be compatible with the plant and animal species of the area, and must meet the requirements of other applicable federal and state laws and regulations, including those that apply to poisonous and noxious plants.[165] OSM also has issued an internal policy directive to promote the use of native species in the reforestation of surface lands disturbed or impacted by coal mining, rather than introduced species.[166]

Department of State

The Department of State helps develop U.S. foreign policy on invasive species, and presents the U.S. position and policies within an international context, including conventions, regional initiatives, and bilateral agreements. Some of these agreements and treaties are discussed in the text box below. The Department of State participates in projects, initiatives, and workshops on invasive species to raise awareness, to share data and information, and also to build regional and global capacity to address invasive species prevention and management. The State Department works with other federal agencies, states, tribes, non-governmental organizations and the private sector. Funding for Department of State activities on invasive species accounted for $18.4 million (less than 1% of total federal funding for invasive species) in FY2012 (Table 2).

Department of Transportation

The Department of Transportation has a policy on invasive species, focusing on coordination and cooperation with other federal agencies as well as with international organizations, such as the International Maritime Organization, the International Civil Aviation Organization, and the International Organization for Standardization to prevent and control the spread of invasive species.[167] Funding for Department of Transportation activities on invasive species accounted for less than $400,000 in FY2012—a relatively small share of total federal funding for invasive species (Table 2).

Federal Aviation Administration

The Federal Aviation Administration (FAA) cooperates with other federal and state agencies to reduce the risk of introducing invasive species at airports, particularly regarding arrivals in Hawaii. For example, FAA works with USDA to enforce the Animal Damage Control Act (7 U.S.C. §§426 *et seq.*) to control for wildlife hazards on or near airports. FAA also cooperates in federal research for screening baggage, cargo, and passengers; and protects native species in the management of its facilities and FAA-funded and licensed facilities throughout the country. In addition, FAA was instrumental in promoting involvement by the International Civil Aviation Organization (ICAO) relative to invasive species. ICAO acknowledged the significance of civil aviation as a pathway for the introduction of invasive species and now urges all member states to use national civil aviation authorities to assist in reducing the risk of introducing invasive species to areas outside their natural range.[168]

Federal Highway Administration

The Federal Highway Administration (FHWA) has an oversight role in federally funded highway projects, including both interstate and state highways, and works with other federal agencies and state governments to prevent the introduction and spread of invasive species. Its vegetation management guides help state departments of transportation prevent the introduction and spread of invasive species.[169] After Executive Order 13112 was issued, FHWA provided guidance to the states, encouraging inventory and integrated management of roadside weeds before-and-after projects. The guidance requires invasive species assessment during the NEPA process. The agency continues to provide technical support to all states on this vegetation issue.

Selected International Efforts Addressing Invasive Species

Executive Order 13112 is the primary authority for the broad coordination of international invasive species activity carried out by most U.S. federal agencies as part of their participation in NISC. In addition, there are a number of international organizations cooperating on matters concerning invasive species; however, the United States is a party to only a very limited number of international treaties or conventions in this area. Some selected U.S. efforts in the international arena include:

- **Convention on International Trade in Endangered Species of Wild Fauna and Flora (CITES), 1973**. A multilateral treaty that regulates the intentional trade in certain species of wildlife and plants worldwide, and calls for the consideration of invasive species when developing national legislation and regulations that deal with the trade in live animals or plants. The United States is a signatory of CITES.
- **South Pacific Regional Environment Program (SPREP) Convention, 1990**. A multilateral treaty that calls on its signatory nations to take measures to protect rare or threatened ecosystems and species within the region. The United States is a member of SPREP.
- **Convention on Biological Diversity (CBD), 1993**. A multilateral treaty that calls upon its signatory nations to prevent the introduction of, control, or eradicate those alien species that threaten ecosystems, habitats, or species, to the extent practicable.

The United States is an observer but is not a signatory of the treaty.

- **World Trade Organization (WTO) Agreement on the Application of Sanitary and Phytosanitary (SPS) Measures, 1995**. The SPS Agreement provides a uniform framework for measures to protect the health and lives of humans, plants, and animals. The agreement recognizes the international standards, guidelines, and recommendations of three organizations—Codex Alimentarius Commission (Codex); International Plant Protection Convention (IPPC), and Office International des Epizooties (OIE).
- Accordingly, Codex aims to protect the health of consumers and has formulated safety standards for specific food products, as well as standards regarding pesticide and drug residues, food contaminants and additives, labeling, and other food safety concerns. IPPC provisions and standards apply to quarantine pests in international trade, while OIE standards are aimed preventing the spread of animal diseases. The United States is a member and active participant within each of the three organizations and the WTO, and a signatory of the SPS Agreement.
- **North American Free Trade Agreement, 1995**. The free trade agreement between the United States, Canada and Mexico, and the side agreement to NAFTA, the North American Agreement for Environmental Cooperation (NAAEC), plays a role in protecting native biota within the three countries. The Commission for Environmental Co-Operation (CEC) under the agreement may consider and develop recommendations regarding non-native species that may be harmful.
- **Code of Conduct for Responsible Fisheries, 1995**. As part of the Food and Agriculture Organization of the United Nations provides for guidelines and recommendations for the responsible use of non-native species in fisheries and aquaculture.
- **International Maritime Organization (IMO) Resolutions, 1991 and 1997**. IMO is a specialized agency of the United Nations and encourages its members apply two sets of guidelines—Resolution A.868 (20) and Resolution A.74 (18)—regarding ballast water exchange that provide guidance and strategies to minimize the risk of invasive species, and other unwanted organisms, from ballast water and sediment discharge.

- **International Civil Aviation Organization (ICAO), Assembly Resolution, 1998**. ICAO is a specialized agency of the United Nations and encourages the aviation authorities of its member nations to assist in the prevention and control of invasive species. International Convention for the Control and Management of Ships' Ballast Water and Sediments, 2004. A multilateral treaty would require the ships of its signatory nations to implement a Ballast Water and Sediments Management Plan and to carry a Ballast Water Record Book and be required to conduct certain procedures.

Though not yet a signatory, the United States was one of the major proponents of this convention.

Sources: Prepared by CRS from various sources including NISC, 2001 *Management Plan National Invasive Species Council*, January 2001, http://www.invasivespecies.gov/main_nav/mn_NISC_ManagementPlan.html.

Federal Railroad Administration

Historically, railroads were a source of invasive species, as livestock were transported around the country and the seeds riding in or on the animals were shed from livestock cars. Now, the Federal Railroad Administration (FRA) works with other federal agencies to support invasive species control efforts on rail corridors. For example, FRA has worked to reduce the risk from invasive species through cooperation with USDA in implementing regulations to lessen the opportunity for spreading karnal bunt, a fungal disease infecting wheat and triticale, across international borders.

Department of the Treasury

The Department of the Treasury advises and assists in the formulation and execution of U.S. international economic and financial policy, including the development of policies with respect to international trade, investment, bilateral aid, environment, and development programs.[170] Previously, multilateral development banks have funded invasive species projects.

Executive Office of the President

Council on Environmental Quality

The Council on Environmental Quality (CEQ) coordinated and led the Clinton Administration's efforts on developing Executive Order 13112 and assisted with formulating agency guidance to implement NEPA. CEQ also chairs the Asian Carp Regional Coordinating Committee in an effort to better protect the Great Lakes from Asian carp (*Hypophthalmichthys* sp.).[171] Several agencies serve on this committee, including FWS, NOAA, EPA, among other agencies.

Office of Science and Technology Policy

The Office of Science and Technology Policy (OSTP) developed a white paper on Executive Order 13112. The Committee on Environment, Natural Resources, and Sustainability (CENRS) of the President's National Science and Technology Council, which is administered through OSTP.[172] CENRS provides for interagency coordination on environment, natural resources, and sustainability policy, and research and development both domestically and internationally. Invasive species are among CENRS research priorities. Representatives from several federal agencies serve on CENRS, including Defense, EPA, Interior, NOAA, National Science Foundation, USDA, the Smithsonian Institution, State, Transportation, and the Council on Environmental Quality. Executive Order 13112 specifically directs NISC to work with CENRS.

Office of the United States Trade Representative (USTR)

The Office of the United States Trade Representative (USTR) has lead responsibility for developing and coordinating U.S. international trade, commodity and direct investment policy. USTR also leads interagency processes to review the potential environmental effects of trade negotiations. It also is the lead trade negotiator for U.S. bilateral, regional and multilateral trade and investment agreements. USTR leads or directs U.S. negotiations with other countries as part of the U.S. membership with the World Trade Organization (WTO) and on multilateral agreements that address certain human, plant, and animal health concerns, such as the Agreement on Sanitary and Phytosanitary (SPS) Measures.[173]

Independent Agencies

Agency for International Development

The U.S. Agency for International Development (USAID) provides economic, development, and humanitarian assistance to developing countries in support of the foreign policy goals of the United States. USAID has responsibility for ensuring that U.S. development assistance programs do not lead to the introduction of invasive species in other nations. Its programs support projects to eradicate and control invasive species where they are already established in developing countries, especially when food, water, and health security are at risk.[174]

Environmental Protection Agency

The Environmental Protection Agency (EPA) is a member of NISC and actively participates in implementing its invasive species management plan. EPA conducts and supports research on the prevention, early detection, control, and management of invasive species. EPA is involved internationally in cooperative efforts focusing on developing policies for early detection and rapid response to potential invasive species. These efforts are mainly coordinated from EPA's regional offices surrounding the Great Lakes and regional offices in the Northeast and involve considerable collaboration and cooperation with Canadian environmental resource managers. As part of these international efforts, EPA is designing public awareness programs on the risks and impacts of invasive species. Invasive species public awareness programs are also sponsored by various estuarine management groups, administered by EPA.

EPA administers a number of environmental laws that have some aspects relating to invasive species, although invasive species may not be the primary focus of these laws. For example, under the Federal Insecticide, Fungicide, and Rodenticide Act (FIFRA, 7 U.S.C. §§136-136y) EPA oversees the federal registration process for pesticides, including pesticides that may be used to control and/or eradicate invasive species, and may place limits on the conditions under which such compounds may be used. EPA implements the Clean Water Act (CWA, 33 U.S.C. §§1251- 1376), which broadly aims to protect waters of the United States. CWA permit programs regulate ballast water discharges and also protect U.S. wetlands and waterways from invasive plants and aquatic species. Under CWA, EPA is promulgating joint regulations with DoD to set uniform national discharge standards and requiring the use of marine pollution control devices to control incidental discharges from armed forces vessels.[175]

EPA also conducts research to develop methodologies for the early detection of non-native invasive species and approaches for applying those methodologies in existing environmental monitoring programs. EPA's research activities include evaluation of ecological indicators (including non-native species) for surface waters, the effects of non-native species on wetland restoration, and studies on non-native submerged aquatic vegetation, as well as model development to estimate the area and spreading rates for potentially invasive species. EPA scientists are also working with regional staff to develop guidance for including evaluations of the potential impacts of invasive species in NEPA assessments.

National Science Foundation

The National Science Foundation (NSF) funds basic and applied research on invasive species, including their roles in population and ecological processes, their relationship to biological conservation activities, and their role as a disturbance agent in ecosystems.[176]

Smithsonian Institution

As part of the broader Smithsonian Institution, the Smithsonian Environmental Research Center (SERC) performs research on invasive species in coastal ecosystems, and collects and analyses data and reports.[177] Specific projects have, for example, examined patterns of ballast-water delivery, tested the susceptibility of marine communities to invasive species, and measured species transfer associated with shipping. In addition, in cooperation with the U.S. Coast Guard, SERC has established the National Ballast Water Information Clearinghouse to measure the changing patterns of ballast water delivery and management for vessels arriving in U.S. ports and to synthesize national data on patterns and impacts of alien species in coastal ecosystems.[178]

SERC has also documented the history of alien species invasions for Chesapeake Bay, among other projects. Aquatic and terrestrial invasive species research is also conducted by the Smithsonian Institution's National Zoological Park, the Smithsonian Tropical Research Institute, and the National Museum of Natural History.

APPENDIX A. MAJOR FEDERAL AGENCIES AND LAWS GOVERNING INVASIVE SPECIES

Agency	Major Responsibilities and Activities	Selected Authorities, as amended
Department of Agriculture		
Animal and Plant Health Inspection Service (APHIS)	Protects U.S. agriculture from domestic and foreign pests and diseases, responds to domestic animal and plant health problems, and facilitates agricultural trade. As part of its regulatory framework, APHIS has oversight of animal and plant health, including the prevention of foreign diseases and pests, eradication and containment of such problems domestically.	As amended: Animal Health Protection Act (7 U.S.C. §§8301-8322); Plant Health Protection Act (7 U.S.C. §§7701-7721); Agricultural Bioterrorism Act (7 U.S.C. §8401); Animal Damage Control Act (7 U.S.C. §§426 *et seq.*); Federal Seed Act (7 U.S.C. §§1551 *et seq.*); Federal Noxious Weed Act (7 U.S.C. §2814); Noxious Weed Control and Eradication Act of 2004 (7 U.S.C. §§7781-7786); National Environmental Policy Act (42 U.S.C. §§4321 *et seq.*); among other authorities,
USDA's Research, Education, and Economics agencies: Agricultural Research Service (ARS), Economic Research Service (ERS), and National Institute of Food and Agriculture (NIFA).	ARS is USDA's chief scientific in-house research agency. Provides scientific and technical support for its regulatory agencies. ERS is USDA's economic research agency, and supports invasive species efforts through its various research programs. NIFA coordinates and administers federal funding of land grant and other institutions to conduct agricultural and food research, and education activities, including research on invasive species.	Numerous laws dating to the Department of Agriculture Organic Act of 1862 (7 U.S.C. §2201 note), up through and including various omnibus farm bill laws.

(Continued)

Agency	Major Responsibilities and Activities	Selected Authorities, as amended
Farm Service Agency (FSA)	In managing the Conservation Reserve Program (CRP), requires all participants to control weeds (including noxious weeds), insects, pests, and other undesirable species on enrolled lands.	Provisions governing CRP (16 U.S.C. §§3838a, 3832), as amended; National Environmental Policy Act (42 U.S.C. §§4321 *et seq.*).
Foreign Agricultural Service (FAS)	Works with APHIS, helps provide invasive species technical assistance to foreign countries.	See laws and statutes under APHIS.
Forest Service (FS)	Manages invasive activities on 193 million acres of national forests and grasslands, as well as supports activities outside the United States.	As amended: Organic Administration Act of 1897 (16 U.S.C. §551); Multiple-Use Sustained-Yield Act (16 U.S.C. §§528-531); Forest and Rangeland Renewable Resources Planning Act (16 U.S.C. §§1671 *et seq.*), as amended by the National Forest Management Act (16 U.S.C. §1604); Federal Noxious Weed Act (7 U.S.C. §2814); Public Rangelands Improvement Act (43 U.S.C. §§1901 *et seq.*); Federal Land Policy and Management Act (43 U.S.C. §1701 *et seq.*); Hawaii Tropical Forest Recovery Act (16 U.S.C. §4503(note)); Cooperative Forestry Assistance Act (16 U.S.C. §§2101-2111); sections of the International Forestry Cooperation Act (16 U.S.C. §4501(b)); National Environmental Policy Act (42 U.S.C. §§4321 *et seq.*); among other laws.

Agency	Major Responsibilities and Activities	Selected Authorities, as amended
Natural Resources Conservation Service (NRCS)	Provides technical assistance to cooperating landowners and federal agencies (such as the Forest Service and Bureau of Land Management) to adopt conservation practices on agricultural land, including rangeland, and promotes conservation planning through many of its farmland conservation programs.	As amended: Soil Conservation and Domestic Allotment Act (16 U.S.C. §590(a)-590(f)); also farmland conservation provisions in various omnibus farm bill laws (e.g., Food Security Act (16 U.S.C. §3839aa-3839aa-8); Federal Agriculture Improvement and Reform Act of 1996 (16 U.S.C. §3836a); National Environmental Policy Act (42 U.S.C. §§4321 *et seq.*).
Department of Commerce		
National Oceanic and Atmospheric Administration (NOAA)	Administers a variety of programs aimed at expanding and coordinating prevention, early detection, rapid response, control, and monitoring programs nationwide, among other roles and responsibilities. NOAA is also the statutory co-chair of both the interagency Aquatic Nuisance Species (ANS) Task Force and National Invasive Species Council (NISC).	As amended: Nonindigenous Aquatic Nuisance Prevention and Control Act (16 U.S.C. §4701, *et seq.*); Endangered Species Act (16 U.S.C. §§1531-1543); Fishery Conservation and Management Act (16 U.S.C. §§1801-1882); Coastal Zone Management Act (16 U.S.C. §§1451 *et seq.*); Interjurisdictional Fisheries Act 1986 (16 U.S.C. §§4101 *et seq.*); National Environmental Policy Act (42 U.S.C. §§4321 *et seq.*).
Department of Defense		
U.S. Army Corps of Engineers (USACE)	Maintains programs that address subcategories of invasive species, and provides guidance and	As amended: Nonindigenous Aquatic Nuisance Prevention and Control Act (16 U.S.C.

(Continued)

Agency	Major Responsibilities and Activities	Selected Authorities, as amended
	research assistance on invasive species control strategies. Provides support to states for aquatic plant management and funds control of invasive aquatic plants in certain southeastern states.	§§4701, *et seq.*), Rivers and Harbors Appropriation Act, as amended (33 U.S.C. §403); Water Resources Development Act (§3061); and National Environmental Policy Act (42 U.S.C. §§4321 *et seq.*).
Department of Health and Human Service (HHS)	Within HHS, the Centers for Disease Control and Prevention (CDC) address zoonotic and emerging diseases. In addition, the National Institute of Health (NIH) supports zoonotic and bioterrorism preparedness research.	Homeland Security Act of 2002 (6 U.S.C. §§101 *et seq.*); sections of the Public Health Service Act (42 U.S.C. §201 *et seq.*) and other public health authorities.
Department of Homeland Security		
U.S. Coast Guard	Responsible for developing and implementing a ballast water management program to prevent the unintentional introduction/dispersal of nonindigenous aquatic species from ship ballast water	As amended: Nonindigenous Aquatic Nuisance Prevention and Control Act (16 U.S.C. §§4701, *et seq.*); Federal Water Pollution Control Act of 1948 (also known as Clean Water Act, 33 U.S.C. §§1251-1376); National Environmental Policy Act (42 U.S.C. §§4321 *et seq.*).
Customs and Border Protection (CBP)	Responsible for border protection and facilitating lawful international trade and travel. Works with other federal agencies to enforce laws prohibiting or limiting the entry of invasive species.	Homeland Security Act of 2002 (6 U.S.C. §§101 *et seq.*); Tariff Act (19 U.S.C. §§1202-1654).

Agency	Major Responsibilities and Activities	Selected Authorities, as amended
Federal Emergency Management Agency's (FEMA)	Works with other federal, state, tribal, and local authorities to control and eradicate outbreaks of animal/zoonotic disease, exotic plant pests, or invasive plant pest infestations; also contributes to the protection of natural and cultural resources.	Homeland Security Act of 2002 (6 U.S.C. §§101 *et seq.*); National Environmental Policy Act (42 U.S.C. §§4321 *et seq.*); among other authorities.
Immigration & Customs Enforcement (ICE)	DHS' principal investigative arm, responsible for border control, customs, trade and immigration.	Homeland Security Act of 2002 (6 U.S.C. §§101 *et seq.*).
Department of the Interior		
Bureau of Indian Affairs (BIA)	Helps support the management of non-native species on Indian lands through its exotic weed eradication and other programs.	As amended: Federal Noxious Weed Act (7 U.S.C. §2814); National Environmental Policy Act (42 U.S.C. §§4321 *et seq.*); among other authorities..
Bureau of Land Management (BLM)	Controls for non-native and invasive plants on land it manages, primarily in western states and Alaska.	Federal Land Policy and Management Act of 1976 (43 U.S.C. §§1701 *et seq.*); National Environmental Policy Act (42 U.S.C. §§4321 *et seq.*); among other authorities.
Bureau of Reclamation (BOR)	Conducts research, prevention, detection, and controls to address pests of aquatic systems such as reservoirs, canals, pipelines, and rivers.	Reclamation Act of 1902 (43 U.S.C. §391h), Fish and Wildlife Coordination Act, as amended (16 U.S.C. §§661-667e; the Act of March 10, 1934; Ch. 55; 48 Stat. 401); National Environmental Policy Act (42 U.S.C. §§4321 *et seq.*).

(Continued)

Agency	Major Responsibilities and Activities	Selected Authorities, as amended
Fish and Wildlife Service (FWS)	Works to conserve, protect, and enhance fish, wildlife, plants and their habitats. Works to prevent the introduction and spread of invasive species, and on controlling established non-native	As amended: Lacey Act (18 U.S.C. §§42-43; 16 U.S.C. §§3371-3378); Endangered Species Act (16 U.S.C. §1531-1543); Nonindigenous Aquatic Nuisance Prevention and Control Act
	species, often working with other agencies (USDA, NOAA, and CBP). Maintains programs covering fisheries, endangered species, habitat conservation, refuge operations and maintenance, and international affairs.	(16 U.S.C. §§4701, *et seq.*); Wild Bird Conservation Act (16 U.S.C. §§4901, *et seq.*); Hawaii Tropical Forest Recovery Act (16 U.S.C. §4503(note)); National Environmental Policy Act (42 U.S.C. §§4321 *et seq.*); among other authorities.
Geological Survey (USGS)	Supports efforts to identify, document, disseminate, and integrate information about the nation's biological resource, including nonindigenous species.	Organic Act of March 3, 1879 (43 U.S.C. 31); Fish and Wildlife Resources Cooperative Agreements (16 U.S.C. §753 a); Agreements to Implement the Convention on Great Lakes Fisheries between the United States and Canada (16 U.S.C. §§931-939); Clean Water Act (33 U.S.C. §§1251-1387); among other authorities.
National Park Service (NPS)	Uses an integrated pest management approach to manage exotic species, and targets specific sites or species. Regulates fishing on its lands and prohibits the possession or use of certain bait for fishing.	As amended: National Park System (16 U.S.C. §§1 *et seq.*; 16 U.S.C. §594); Endangered Species Act (16 U.S.C. §1531 *et seq.*); Noxious Weed Control and Eradication Act (7 U.S.C. §§7781-7786); Plant Protection Act (7 U.S.C. §7701 *et seq.*); National Invasive

Agency	Major Responsibilities and Activities	Selected Authorities, as amended
		Species Act (16 U.S.C. 4701); Nonindigenous Aquatic Nuisance Prevention and Control Act (16 U.S.C. §§4701); Animal Damage Control Act (7 U.S.C. §§426-426c); and National Environmental Policy Act (42 U.S.C. §§4321 *et seq.*); among other authorities.
Office of Surface Mining Reclamation and Enforcement (OSM)	Addresses the use of introduced species in mine reclamation areas for revegetation of impacted lands through regulations governing coal mining operations.	Regulations implementing the Surface Mining Control and Reclamation Act (30 U.S.C. §§1201 *et seq.*).
Department of State	Works with other federal agencies, states, tribes, non-governmental organizations and industry to develop U.S. foreign policy on invasive species.	State Department Basic Authorities Act (22 U.S.C. §2651a), as well as 1 U.S.C. §§U.S.C. 112a, 1 U.S.C. 112b, 1 U.S.C. 113 112b(c), among other domestic and international legal authorities.
Department of Transportation		
Federal Aviation Administration (FAA)	Cooperates with other federal and state agencies to develop a strategy to reduce the risk of introducing invasive species at airports. Protects native species at its facilities and in FAA-funded and licensed facilities.	Various environmental laws such as Endangered Species Act of 1973 (16 U.S.C. §1531 *et seq.*); Clean Water Act (33 U.S.C. §§1251-1387); Animal Damage Control Act (7 U.S.C. §§426 *et seq.*); National Environmental Policy Act (42 U.S.C. §§4321 *et seq.*);
Federal Highway Administration (FHWA)	Works with other federal agencies and state governments to combat the introduction and spread of invasive species.	Various environmental laws such as Endangered Species Act of 1973 (16 U.S.C. §1531 *et seq.*); Clean Water Act (33 U.S.C. §§1251-1387);

(Continued)

Agency	Major Responsibilities and Activities	Selected Authorities, as amended
		National Environmental Policy Act (42 U.S.C. §§4321 *et seq.*)
Federal Railroad Administration (FRA)	Works with other federal agencies to support invasive species control efforts on rail corridors.	Various environmental laws such as Endangered Species Act of 1973 (16 U.S.C. §1531 *et seq.*); Clean Water Act (33 U.S.C. §§1251-1387); National Environmental Policy Act (42 U.S.C. §§4321 *et seq.*)
U.S Agency for International Development (USAID)	Responsible for ensuring that U.S. development of assistance programs do not lead to the introduction of invasive species in other nations, and supports eradication and control of invasive species in developing countries.	Foreign Assistance Act of 1961, as amended (22 U.S.C. §§2347 *et seq.*), among other legal authorities.
U.S. Environmental Protection Agency (EPA)	Assists with cooperative efforts regarding early detection and rapid response to potential invasive species. Conducts and supports research on the prevention, early detection, control, and management of invasive species. Responsible for establishing numeric limits on organisms in ballast water discharges, as well as additional monitoring and reporting of vessel discharges.	Clean Water Act (33 U.S.C. §§1251-1387); National Environmental Policy Act (42 U.S.C. §§4321 *et seq.*), among other environmental laws and authorities.

Source: Prepared by CRS from various sources including National Invasive Species Information Center (NISIC), "National Management Plan: Survey of Federal Roles and Responsibilities," August 2009, http://www.invasivespeciesinfo.gov/council/ survey.shtml; and NISC, 2001 *Management Plan National Invasive Species Council*, January 2001, http://www.invasivespecies.gov/main_nav/mn_NISC_ ManagementPlan.html. Other source information is from individual agency websites.

APPENDIX B. LIST OF ACRONYMS

AHPA	Animal Health Protection Act
APHIS	Animal and Plant Health Inspection Service
ARS	Agricultural Research Service
ASPEA	Alien Species Prevention and Enforcement Act
BIA	Bureau of Indian Affairs
BLM	Bureau of Land Management
BOR	Bureau of Reclamation
BWM	Ballast water management
CEQ	Council on Environmental Quality
CENRS	Committee on Environment, Natural Resources, and Sustainability
DOD	Department of Defense
DOI	Department of the Interior
EPA	Environmental Protection Agency
ERS	Economic Research Service
FAA	Federal Aviation Administration
FHW	A Federal Highway Administration
FICMNEW	Federal Interagency Committee for Management of Noxious and Exotic Weeds
FSA	Farm Service Agency
FWS	Fish and Wildlife Service
GLERL	Great Lakes Environmental Research Lab
NANPCA	Non-indigenous Aquatic Nuisance Prevention and Control Act
NEPA	National Environmental Policy Act
NIFA	National Institute of Food and Agriculture
NISA	National Invasive Species Act
NISC	National Invasive Species Council
NMFS	National Marine Fisheries Service
NOAA	National Oceanic and Atmospheric Administration
NPS	National Park Service
NRCS	Natural Resources Conservation Service
NWRS	National Wildlife Refuge System
OSM	Office of Surface Mining Reclamation and Enforcement
OSTP	Office of Science and Technology Policy
PPA	Plant Protection Act
USACE	U.S. Army Corp of Engineers

USGS	U.S. Geological Survey
USTR	United States Trade Representative
WBCA	Wild Bird Conservation Act

End Notes

[1] Despite differences in emphasis, this report generally uses these terms interchangeably.

[2] D. Pimentel, "Environmental and Economic Costs Associated with Alien Invasive Species in the United States," in *Biological Invasions: Economic and Environmental Costs of Alien Plant, Animal, and Microbe Species*, CRC Press, 2011. Updates estimates from previous studies in 2000 and 2005, and the first edition of the book.

[3] The University of Florida's Center for Aquatic and Invasive Plants provides a compilation of the available lists, laws, rules, and regulations by state and state agencies addressing non-native plants (http://plants.ifas.ufl.edu/node/634).

[4] D. Pimentel, "Environmental and Economic Costs Associated with Alien Invasive Species in the United States," in *Biological Invasions: Economic and Environmental Costs of Alien Plant, Animal, and Microbe Species*, CRC Press, 2011. Updates previous estimates from previous studies in 2000 and 2005, and the first edition of the book.

[5] This report excludes costs associated with certain human diseases, estimated at nearly $92 billion annually. Human disease costs include reportedly introduced diseases such as AIDS, syphilis, and influenza.

[6] See, for example, L. Goldstein, "17 Reasons the Economic Impact of the Domestic Cat as a Non-Native Species in the U.S. Does Not Cost $17 Billion," 2011; and D. Gattuso, "Invasive Species: Animal, Vegetable or Political?" National Center for Public Policy Research, August 2006.

[7] See, for example, S. Lovell et al., "The Economic Impacts of Aquatic Invasive Species: A Review of the Literature," *Agricultural and Resource Economics Review*, 35 (1):195–209, April 2006; and C. Zhang and K. Boyle, "The Effect of an Aquatic Invasive Species (*Eurasian watermilfoil*) on Lakefront Property Values," *Ecological Economics*, Vol. 70 (2): 394-404, December 2010.

[8] See, for example, National Invasive Species Information Center (NISIC), "National Management Plan: Survey of Federal Roles and Responsibilities," August 2009, http://www.invasivespeciesinfo.gov/council/survey.shtml; and NISC, 2001 *Management Plan National Invasive Species Council*, January 2001, http://www.invasivespecies.gov/main_nav/mn_NISC_ManagementPlan.html.

[9] Federal Insecticide, Fungicide, and Rodenticide Act (FIFRA, 7 U.S.C. §§136-136y); Federal Water Pollution Control Act of 1948, also known as Clean Water Act (CWA, 33 U.S.C. §§1251-1376). For example, FIFRA regulations broadly govern pesticide use, including pesticides that may be used to control and/or eradicate invasive plants; CWA broadly aims to protect waters of the United States, and contains provisions regulating ballast water discharges and also protecting U.S. wetlands and waterways from invasive plants and aquatic species.

[10] Fish and Wildlife Act of 1956 (16 U.S.C. §742).

[11] For more information, see CRS Report RL33152, *The National Environmental Policy Act (NEPA): Background and Implementation*.

[12] This section was adapted in part from CRS Report R43170, *Injurious Species Listings Under the Lacey Act: A LegalBriefing*.

[13] Plants are not covered by the injurious species provision.

[14] The trade provisions are in the conservation title (Title 16) of the U.S. Code. For more information in this aspect of the Lacey Act, see CRS Report RL34395, *International Illegal Trade in Wildlife: Threats and U.S. Policy*; and CRS Report R42067, *The Lacey Act: Protecting the Environment by Restricting Trade*.

[15] In addition, the regulations also ban certain dead fish.

[16] Although this provision of Title 18 of U.S. Code makes no mention of insects and spiders, provisions in Title 16 do cover these species. There are many other examples of differences in coverage between the act's major provisions.

[17] FWS, "Species Listed as Injurious Wildlife under the Lacey Act (50 CFR 16)," http://www.fws.gov/injuriouswildlife/pdf_files/Current_Listed_IW.pdf.

[18] Generally, federal laws have tended to focus on exclusion, or "black lists," i.e., on species that have already been shown to be harmful (anything not on the list is allowed), in contrast to a "white list" (anything not on the list is excluded). See, for example, the Federal Noxious Weed List (http://plants.usda.gov/) where the U.S. government has designated certain plants as noxious weeds, in accordance with the Plant Protection Act (7 U.S.C. §§7701 *et seq*.).

[19] 31 Stat. 188. These four were not all accompanied by scientific names in the original statute. Thus, the number of species actually banned in 1900 may have been as high 90. The 1949 amendments also added language prohibiting transport of wild animals and birds "under inhumane and unhealthful conditions." 63 Stat. 89. While this section remains codified with the injurious species provision in 18 U.S.C. §42(c), it is not part of the ban on injurious species.

[20] The 1949 amendments removed language that barred import of "any foreign wild animal or bird" (63 Stat. 89).

[21] For more information, see CRS Report R42067, *The Lacey Act: Protecting the Environment by Restricting Trade*; and CRS Report R42119, *The Lacey Act: Compliance Issues Related to Importing Plants and Plant Products*.

[22] NISC, *2001 Management Plan National Invasive Species Council*, January 2001, http://www.invasivespecies.gov/main_nav/mn_NISC_ManagementPlan.html.

[23] MOU 225-05-7000 (APHIS Agreement 04-9100-0859-MU) is at http://www.fda.gov/AboutFDA/PartnershipsCollaborations/MemorandaofUnderstandingMOUs/DomesticMOUs/ucm359217.htm.

[24] USDA's "Wildlife Damage Management" website (http://www.aphis.usda.gov/wildlife_damage/). This program was known until 1997 as the Animal Damage Control program.

[25] APHIS's Wildlife Damage Management website: http://www.aphis.usda.gov/wildlife_damage/prog_data/prog_data_report.shtml.

[26] These programs were originally authorized in various omnibus farm bills, including provisions in the Food Security Act of 1985 (16 U.S.C. §§3839aa-3839aa-8) and the Federal Agriculture Improvement and Reform Act of 1996 (16 U.S.C. §3836a).

[27] NISC, *2001 Management Plan National Invasive Species Council*, January 2001, http://www.invasivespecies.gov/main_nav/mn_NISC_ManagementPlan.html.

[28] This section was prepared by Linda Luther, Analyst in Environmental Policy (lluther@crs.loc.gov, 7-6852).

[29] For an overview of NEPA, see CRS Report RL33152, *The National Environmental Policy Act (NEPA): Background and Implementation*.

[30] 42 U.S.C. §4332(2)(C). Required elements of an EIS are specified in the Council on Environmental Quality (CEQ) regulations implementing NEPA (40 CFR Parts 1500-1508) and broadly apply to all federal agencies.

[31] 40 CFR Part 1508.18; this definition also specifies that new or revised agency rules, regulations, plans, policies, or procedures are subject to NEPA.

[32] Depending on an action's impacts and the agency implementing it, an action may be subject to various laws, regulations, executive orders, or agency-specific requirements. In addition to federal requirements, the project also may be subject to local, state, or tribal laws or regulations.

[33] For information about the ESA generally, see CRS Report RL31654, *The Endangered Species Act: A Primer.*
[34] 16 U.S.C. §1536.
[35] For information on ESA's consultation process, see CRS Report RL31654, *The Endangered Species Act: A Primer.*
[36] 16 U.S.C. §1539.
[37] 7 U.S.C. §§7701 *et seq.*
[38] P.L. 101-624, Title XIV (amending the original law, P.L. 93-629).
[39] APHIS's Noxious Weeds Program website: http://www.aphis.usda.gov/plant_health/plant_pest_info/weeds/index.shtml.
[40] Provisions pertaining to range improvements in the Act of April 24, 1950 (16 U.S.C. §580(h)), also state that of funds received from grazing fees, a portion can be used for the "eradication of poisonous plants and noxious weeds in order to protect or improve the future productivity of the range."
[41] USDA, "Forest Service Authorities for Invasive Species Management," March 2006, http://www.fs.fed.us/invasivespecies/policy.shtml.
[42] NANPCA, §1101 (under U.S. Coast Guard regulations at 33 CFR Part 151).
[43] For more information on ANSTF, see report section "Aquatic Nuisance Species Task Force."
[44] In addition, other NOAA activities fall under other authorities, including control of phragmites under essential fish habitat provisions of the Magnuson-Stevens Fishery Conservation and Management Act of 1976 (16 U.S.C. §§1801-1882) and other activities under the Coastal Zone Management Act of 1972 (16 U.S.C. §§1451 *et seq.*) and the Interjurisdictional Fisheries Act 1986 (16 U.S.C. §§4101 *et seq.*), as amended.
[45] NISC, *2001 National Invasive Species Management Plan*, January 2001, http://www.invasivespecies.gov/main_nav/mn_NISC_ManagementPlan.html.
[46] Grants were authorized to states for implementing approved management plans, with maximum federal shares of 75% of costs for each comprehensive management plan, and 50% for each public facility management plan. However, despite substantial authorizations, relatively little was appropriated or made available for state grants to implement these management plans.
[47] Subsequently, the Corps established the Zebra Mussel Research Program, including annual technical conferences and a publication series (see http://el.erdc.usace.army.mil/zebra/zebra.html).
[48] 33 U.S.C. §§1121 et seq., as amended. More information is at http://seagrant.noaa.gov/Home.aspx.
[49] The Sea Grant program also manages an Aquatic Nuisance Species Clearinghouse website at http://newyorkinvasivespecies.info/ai/default.aspx.
[50] NANPCA, as amended by P.L. 104-332. For more information, see CRS Report RL32344, *Ballast Water Management to Combat Invasive Species.*
[51] This national program became mandatory in 1999 (64 *Federal Register* 94: 26672-26690, May 17, 1999). The U.S. Coast Guard finalized quantitative standards for ballast water treatment in 2012 (77 *Federal Register* 57: 17254-17320, March 23, 2012). For more information, see the report section "Coast Guard."
[52] Before passage of this act, the United States was the leading importer of wild birds; after passage, the European Union displaced the U.S. as the leading importer. See "International Trade in Wild Birds, and Related Bird Movements, in Latin America and the Caribbean," Food and Agriculture Organization of the United Nations, 2008, http://www.rlc.fao.org/es/prioridades/transfron/aviar/pdf/birtr.pdf.
[53] Activities under the Hawaiian biological control program of the Forest Service are described at http://www.fs.fed.us/psw/topics/biocontrol/strawberryguava/biocontrol.shtml.
[54] The plan is available at http://www.state.hi.us/dlnr/dofaw/pubs/HITropicalForestRecoveryPlan.pdf.

[55] The PPA became law in June 2000 as part of the Agricultural Risk Protection Act (P.L. 106-224, Title IV), and consolidated and superseded several U.S. plant health laws, including (1) The Act of August 20, 1912 (commonly known as the "Plant Quarantine Act", 7 U.S.C. §§151-164a, 167); (2) The Federal Plant Pest Act (7 U.S.C. §§150aa *et seq.* and 7 U.S.C. §147a); (3) Section 102 (a) - (e) of the Department of Agriculture Organic Act of 1944 (7 U.S.C. §147a); (4) The Federal Noxious Weed Act of 1974 (7 U.S.C. §§2801 *et seq.*), except sections 1-15 of that act (7 U.S.C. §2801 note and 7 U.S.C. §2814); (5) the Joint Resolution of April 6, 1937 (commonly known as the "Insect Control Act") (7 U.S.C. §§148 *et seq.*); (6) The Halogeton Glomeratus Act (7 U.S.C. §§1651 *et seq.*); (7) The Golden Nematode Act (7 U.S.C. §§150 *et seq.*); and (8) Section 1773 of the Food Security Act of 1985 (P.L. 99-198; 7 U.S.C. §148f).

[56] In fact, USDA has less regulatory authority to address established and widespread pests (i.e., already introduced and established pests) under PPA.

[57] USDA's CCC is a government-owned corporation that is authorized to borrow up to $30 billion at any one time from the U.S. Treasury. The CCC mainly is a financing mechanism for farm bill programs such as commodity price and income supports, agricultural conservation, export assistance, and other mandated authorizations.

[58] 7 U.S.C. §§7751 and 7772. Such cases often occur in response to larger-scale plant and animal pest and disease outbreaks where the costs are too large to cover within existing appropriations or to new and emerging agricultural issues that warrant a federal role. In FY2010, for example, APHIS requested the transfer of emergency funding to address Asian longhorned beetle in Massachusetts ($41.451 million), and to treat potentially large outbreaks of grasshoppers in western states ($10.734 million). For other information, see CRS Report RL32504, *Funding Plant and Animal Health Emergencies: Transfers from the Commodity Credit Corporation.*

[59] 7 U.S.C. §7715.

[60] A summary of USDA quarantine notices is compiled regularly by the National Plant Board and posted on their website (http://nationalplantboard.org/laws/spro.html), organized by pests that trigger a quarantine and the dates of those notices. USDA's website information (http://www.aphis.usda.gov/plant_health/plant_pest_info/) is arranged by pest.

[61] 7 U.S.C. §7756. Procedures on how to process a special needs request are at APHIS, "USDA Establishes Process for States to Impose Additional Restrictions," http://www.aphis.usda.gov/plant_health/special_needs_request/index.shtml.

[62] 7 U.S.C. §7711.

[63] Incorporated laws include USDA's animal quarantine laws across several previous statutes (21 U.S.C. §§101 through 135b and 19 U.S.C. §1306), among others.

[64] This provision is in accordance with the Animal Health Protection Act (7 U.S.C. §§8310 and 8316) and the Plant Protection Act (7 U.S.C. §§7751 and 7772).

[65] 70 *Federal Register* 192:57993-57994. October 5, 2005.

[66] 64 *Federal Register* 6183, February 8, 1999. Executive Order 13112 revoked former President Carter's 1977 Executive Order 11987 on exotic species. For more information, see USDA's website: http://www.invasivespeciesinfo.gov/laws/execorder.shtml. For more information on the legal status of Executive Orders, see CRS Report RS20846, *Executive Orders: Issuance, Modification, and Revocation.*

[67] Information on the roles of federal and independent agencies was, in part, excerpted from the National Invasive Species Council's (NISC) website (http://www.invasivespecies.gov/main_nav/mn_about.html).

[68] See, for example, National Invasive Species Information Center (NISIC), "National Management Plan: Survey of Federal Roles and Responsibilities," August 2009, http://www.invasivespeciesinfo.gov/council/survey.shtml; and NISC, 2001 *Management Plan National Invasive Species Council*, January 2001, http://www.invasivespecies.gov/main_nav/mn_NISC_ManagementPlan.html.

[69] For crosscut budget data for FY2010-FY2012, see "Invasive Species Interagency Crosscut Budget," March 2012, http://www.invasivespecies.gov/global/org_collab_budget/org_collab_budget_documents/ NISC%20FY2011%20Crosscut%20Budget%20Summary.pdf.

[70] For more information, see NISC's website is at http://www.invasivespecies.gov/.

[71] ISAC is a group of nonfederal experts and stakeholders established and mandated by Executive Order 13112 to provide advice to NISC on invasive species-related issues. The advisory committee includes academics, representatives of state and local governments; port authorities; the pet, nursery, and pesticide industries; several environmental groups; a commercial fisherman; and a rancher.

[72] From NISC's website: http://www.invasivespecies.gov/global/prevention/prevention_index.html.

[73] NISC, *2001 National Invasive Species Management Plan*, January 2001, http://www.invasivespecies.gov/main_nav/mn_NISC_ManagementPlan.html.

[74] A five year review is required by Executive Order 13112. NISC, *Five-Year Review of Executive Order 13112 on Invasive Species*, 2005, http://www.invasivespecies.gov/home_documents/Five-Year%20Review-FINAL%20PRINT%20VERSION.pdf. Prepared for and submitted to the Office of Management and Budget.

[75] NISIC develops and manages the www.invasivespeciesinfo.gov website.

[76] NISC, *2008 – 2012 National Invasive Species Management Plan*, August 2008, http://www.invasivespecies.gov/main_nav/mn_NISC_ManagementPlan.html.

[77] NISC's management plan, while discussing improvements in international cooperation in general terms, does not describe any specific agency tasks which the United States itself might identify to prevent the spread of U.S. species that might cause harm in other countries.

[78] 16 U.S.C. §§4701, *et seq.*, as amended by the National Invasive Species Act of 1996.

[79] A list of ANSTF members is at http://anstaskforce.gov/members.php.

[80] State and interstate plans are available at http://anstaskforce.gov/stateplans.php

[81] See "Protect Your Waters" website at http://www.protectyourwaters.net/.

[82] For more information, see FICMNEW's website: http://www.fs.fed.us/ficmnew/index.shtml.

[83] Publications are available at http://www.fs.fed.us/ficmnew/guidance.shtml.

[84] Based on information on FICMNEW's website at http://www.fs.fed.us/ficmnew/background.shtml.

[85] R. M. Nowerierski and H. J. Meyer, "Establishing Inter-agency Multidisciplinary Areawide Pest Management Programmes," *Areawide Pest Management: Theory and Implementation*, ed. O. Koul et al., 2008. For more information, see ITAP's website: http://www.itap.gov/.

[86] Information provided here of the activities at each USDA agency is excerpted from "Avoiding Harm From Invasive Species" report (also known as "USDA Do No Harm Report"), particularly its 2011 report, available at http://www.invasivespeciesinfo.gov/docs/resources/usdanoharm2011.doc.

[87] USDA does not publish a single budget line item for spending on invasive species, in part because USDA's efforts to address invasive species are integral to its mission and part of most of their programs.

[88] Approximated using data from APHIS, compared to total USDA spending reported by NISC in the same year.

[89] Approximated using Forest Service data, compared to total USDA spending reported by NISC in the same year.

[90] NISC, "U.S. Department of Agriculture," http://www.invasivespecies.gov/main_nav/usda.html.

[91] AHPA (7 U.S.C. §§8301-8302); regulations are at 9 CFR Parts 91 through 99.

[92] PPA (7 U.S.C. §§7701 *et seq.*); regulations are at 7 CFR Part 319. The requirements apply to many commodities, including nursery stock.

[93] 7 U.S.C. §2814; regulations are at 7 CFR Parts 360 and 361. Refers to plants that can directly or indirectly injure or cause damage to crops, livestock, poultry, or other interests of agriculture, irrigation, navigation, the natural resources, public health, or the environment.

[94] APHIS, Plant Pest Program Information, http://www.aphis.usda.gov/plant_health/plant_pest_info/. Also see "Federal Noxious Weeds List" (http://www.aphis.usda.gov/plant_health/plant_pest_info/weeds/downloads/weedlist.pdf).
[95] APHIS, Wildlife Services "Invasive Species", http://www.aphis.usda.gov/ wildlife_damage/invasive_species.shtml.
[96] APHIS's Wildlife Damage Management website: http://www.aphis.usda.gov/wildlife_damage/prog_data/prog_data_report.shtml.
[97] 16 U.S.C. §3831. Also see CRS Report R42783, *Conservation Reserve Program (CRP): Status and Issues*.
[98] See, e.g., FSA's Highly Erodible Land Initiative http://www.fsa.usda.gov/Internet/FSA_File/helijobsheet.pdf.
[99] In FY2011, the NFS treated over 352,090 acres of lands and waters infested with invasive species, of which approximately 70,339 acres targeted non-plant invasive species.
[100] USDA, "Avoiding Harm From Invasive Species," various years.
[101] A listing of U.S. Forest Service authorities for invasive species management is at USDA website: http://www.fs.fed.us/invasivespecies/policy.shtml/.
[102] Another USDA data and research agency, the National Agricultural Statistics Service (NASS) is not discussed here.
[103] USDA, "Avoiding Harm From Invasive Species," various years, including USDA, "Avoiding Harm From Invasive Species In FY2011" February 27, 2012, http://www.invasivespeciesinfo.gov/docs/resources/usdanoharm2011.doc.
[104] 42 U.S.C. §§4321 *et seq.*
[105] Section 15 of the Federal Noxious Weed Act, and Executive Order 13112 direct Federal agencies to use an IPM approach for the management of undesirable plants on federal lands, including education; preventive measures; cultural, mechanical, physical, biological and chemical control; and general land management practices (such as revegetation, manipulation of livestock or wildlife grazing, and improvement of livestock and wildlife habitat).
[106] Awarded to 45 recipients, including universities, other USDA agencies, and private nonprofit institutions.
[107] ERS, Program of Research on the Economics of Invasive Species Management Fiscal 2003-2011 Activities, May 2011, http://www.ers.usda.gov/publications/AP/AP056/.
[108] FQPA, enacted in 1996 (P.L. 104-170), amended Federal Insecticide, Fungicide, and Rodenticide Act (FIFRA) and the Federal Food, Drug, and Cosmetic Act (FFDCA). For more information see CRS Report RL31921, *Pesticide Law: A Summary of the Statutes*.
[109] Formerly Cooperative State Research, Education, and Extension Service or CSREES.
[110] USDA, "USDA Grant and Partnership Programs that can Address Invasive Species Research, Technical Assistance, Prevention and Control Federal Fiscal Year 2012," November 2011, http://www.invasivespeciesinfo.gov/docs/toolkit/ usdagrants2012.pdf.
[111] Led by five regional laboratories (Cornell University, University of Florida, Michigan State University, Kansas State University, and University of California-Davis) and one support laboratory (Texas Tech University).
[112] For more information, see USDA's websites: http://www.aphis.usda.gov/animal_health/nahln/downloads/NAHLNBriefingCurrent.pdf and http://www.nifa.usda.gov/nea/ag_biosecurity/in_focus/apb_if_healthlab.html.
[113] For more information, see USDA's website (http://www.aphis.usda.gov/plant_health/permits/tag/index.shtml).
[114] USDA, "Avoiding Harm From Invasive Species," various years.
[115] Programs include Conservation Reserve Program (CRP), Wetlands Reserve Program (WRP), the Wildlife Habitat Incentives Program (WHIP), Environmental Quality Incentives Program (EQIP), and the Conservation Security Program (CSP).
[116] Some of the target weeds in this effort are yellow starthistle, cheatgrass, knapweed, Canada thistle, and cogongrass.

[117] USDA's website: http://www.nrcs.usda.gov/wps/portal/nrcs/main/national/plantsanimals/plants/centers. Centers are located in AK, AZ, CA, CO, FL, GA, HI, ID, KS, LA, MI, MO, MS, MT, ND, NJ, NY, OR, TX, WA, and WV.

[118] The PLANTS database has over 650 fact sheets on-line.

[119] Available at http://directives.sc.egov.usda.gov/ViewRollUp.aspx?hid=17018.

[120] For more information on NOAA's invasive species programs, see http://www.habitat.noaa.gov/restoration/programs/invasivespecies.html.

[121] The Sea Grant Program manages an Aquatic Nuisance Species Clearinghouse website at http://newyorkinvasivespecies.info/ai/default.aspx.

[122] For more information on GLERL's invasive species program, see http://www.glerl.noaa.gov/.

[123] §312(n) of the Clean Water Act (33 U.S.C. §§1251-1376. Regulations are at 40 CFR Part 1700. These standards are referred to as Uniform National Discharge Standards (UNDS). For more information, see EPA's website (http://water.epa.gov/lawsregs/lawsguidance/cwa/vessel/unds/FAQs.cfm).

[124] This section was prepared by Charles Stern, Specialist in Natural Resources Policy (cstern@crs.loc.gov, 7-7786).

[125] U.S. Army Corps of Engineers, Memorandum, *U.S. Army Corps of Engineers Invasive Species Policy*, June 2, 2009. Available at http://corpslakes.usace.army.mil/employees/cecwon/pdfs/09jun12-InvSpeciesRevision.pdf.

[126] 33 U.S.C. §610. Nonfederal entities must provide for 50% of these costs. Corps implementation guidance specifies that maintenance of aquatic plant control at Corps or federally owned projects should be funded through the normal project operations and maintenance budget process. See EP1130-2-500, available at http://publications.usace.army.mil/publications/eng-pamphlets/EP_1130-2-500/c-14.pdf.

[127] This program is fully federally funded and was authorized under NANPCA (16 U.S.C. §§4701, *et seq.*). The laboratory also administers a zebra mussel research effort to develop control measures.

[128] This program was authorized in the Rivers and Harbors Appropriation Act of 1899 (33 U.S.C. §403), as amended.

[129] For more information, see CRS Report R41082, *Asian Carp and the Great Lakes Region*.

[130] Water Resources Development Act (WRDA), as reauthorized in 2007 (P.L. 110-114, §3061), consolidated the multiple authorizations for barrier construction and authorized the Corps to permanently operate both barriers at full federal cost.

[131] The study is authorized in 121 Stat. 586 and is expected to focus on options including the permanent hydrological separation of the basins. For more information, see http://glmris.anl.gov/index.cfm.

[132] Refers to a disease that may be transmitted from animals to humans under natural conditions (e.g., rabies).

[133] A listing of selected authorities pertaining to CDC and public health is at CDC, "Selected Federal Legal Authorities Pertinent to Public Health Emergencies," September 2009, http://www.cdc.gov/phlp/docs/ph-emergencies.pdf.

[134] NISC, "Department of Health and Human Service (HHS)," http://www.invasivespecies.gov/main_nav/hhs.html.

[135] 64 *Federal Register* 94: 26672-26690, May 17, 1999. These regulations became effective July 1, 1999.

[136] Prior to the release of the final rule, the Coast Guard, in a Memorandum of Understanding with EPA, published a report describing generic procedures for evaluating the performance of ballast water treatment systems. EPA, *Generic Protocol for the Verification of Ballast Water Treatment Technology*, EPA/600/R-10/146, September 2010, http://www.uscg.mil/hq/cg5/cg522/cg5224/docs/600r10146.pdf.

[137] 77 Federal Register 57: 17254-17320, March 23, 2012. As required under NANPCA, 16 U.S.C. 4711(c)(2)(A), 16 U.S.C. 4711(c)(2)(E); and 16 U.S.C. 4711(c)(2)(D)(iii).

Regulations are at 33 CFR Part 151, Subparts C and D. For more information, see CRS Report RL32344, *Ballast Water Management to Combat Invasive Species*.

[138] CBP, "Protecting Our Borders," http://www.cbp.gov/xp/cgov/about/mission/cbp.xml. Each year, more than 11 million maritime containers arrive at our seaports. At land borders, another 11 million containers arrive by truck and 2.7 million by rail.

[139] U.S. Fish and Wildlife Service, Office of Law Enforcement, "Strategic Plan 2011 – 2015: Protecting Our Wildlife and Plant Resources," http://www.fws.gov/le/pdf/OLE-Strategic-Plan.pdf, p. 28.

[140] NISC, "U.S. Department of Homeland Security," http://invasivespecies.gov/main_nav/dhs.html.

[141] See, for example, CBP, "On a Typical Day in Fiscal Year 2012" http://www.cbp.gov/linkhandler/cgov/about/accomplish/typical_day_fy12.ctt/typical_day_fy12.pdf.

[142] Emergency Support Function (ESF #11)

[143] ICE, "Overview," http://www.ice.gov/about/overview/.

[144] U.S. Fish and Wildlife Service, Office of Law Enforcement, "Strategic Plan 2011 – 2015: Protecting Our Wildlife and Plant Resources," http://www.fws.gov/le/pdf/OLE-Strategic-Plan.pdf, p. 28, p. 17.

[145] Including American Indian Agricultural Resource Management Act (AIARM, P.L. 103-177, codified at 25 U.S.C. §3701), Federal Noxious Weed Act (7 U.S.C. §2814), and the Federal Insecticide, Fungicide and Rodenticide Act (FIFRA, 7 U.S.C. §§136 *et seq.*).

[146] BLM, *Pulling Together: National Strategy of Invasive Plant Management*, 1998, http://www.blm.gov/wo/st/en/prog/more/weeds.html.

[147] BLM, Invasive Species ("Homework Helpers"), http://www.blm.gov/wo/st/en/res/Education_in_BLM/Learning_Landscapes/For_Kids/homework_helpers/invasive_species.html.

[148] For more on management of wild horses and burros generally, see CRS Report RL34690, *Wild Horses and Burros: Issues and Proposals*.

[149] This section was prepared by Betsy A. Cody, Specialist in Natural Resources Policy (bcody@crs.loc.gov, 7-7229).

[150] Based on its FY2014 budget request for its science and technology program. For more information see also http://www.usbr.gov/mussels/ and http://www.usbr.gov/pmts/eco_research/eco1.html.

[151] Information on the FWS invasive species programs can be found at http://www.fws.gov/invasives/.

[152] 16 U.S.C. §§1531 *et seq.*

[153] USGS, "Invasive Species Program," http://www.usgs.gov/ecosystems/invasive_species/index.html.

[154] See, for example, USGS, "NBII Invasive Species Information Node Information," http://www.fort.usgs.gov/ Research/research_tasks.asp?TaskID=2119.

[155] This section was prepared by Laura B. Comay, Analyst in Natural Resources Policy (lcomay@crs.loc.gov, 7-6036).

[156] NPS, "Invasive Species.....What are they and why are they a problem?" http://www.nature.nps.gov/biology/invasivespecies/.

[157] See http://www.nature.nps.gov/biology/ipm/index.cfm for information on integrated pest management, which includes management of either native or non-native species that interfere with the purpose of a park or threaten human health or safety.

[158] NPS, "Water Resources," http://www.nature.nps.gov/water/marineinvasives/lionfish.cfm.

[159] 36 CFR Part 7.9.

[160] 36 CFR Part 2.3(d)(2). For an example of fishing rules that limit bait types, see those at Cuyahoga Valley National Park: http://www.nps.gov/cuva/planyourvisit/fishing.htm.

[161] For more on the teams, see http://www.nature.nps.gov/biology/invasivespecies/EPMT_teams.cfm. It appears that no similar teams function to control invasion of exotic animals, such as lake trout (*Salvelinus namaycush*) at Yellowstone Lake; animal invasions

continue to be handled on an *ad hoc* basis. However, in its FY2014 budget request, the agency requested increased funding for management of exotic invasive animals, which would "support applied research for development of a science-based service-wide framework that cohesively and effectively addresses invasive terrestrial and aquatic animal issues." See NPS, *Budget Justifications and Performance Information*, Fiscal Year 2014, p. Overview-25, http://www.nps.gov/aboutus/upload/FY_2014_greenbook.pdf.

[162] NPS, "Exotic Plant Management Team Program: 2011 Annual Report," June 2012, http://www.nature.nps.gov/biology/invasivespecies/Documents/EPMT_FY2011_AnnRepor t.pdf, p.2.

[163] This section was prepared by David Bearden, Specialist in Environmental Policy (dbearden@crs.loc.gov, 7-2390).

[164] 30 U.S.C. §§1201 *et seq.*

[165] 30 C.F.R. 816.111 specifies these requirements for the reclamation of lands disturbed by surface coal mining, and 30 C.F.R. 817.111 specifies these requirements for the reclamation of surface impacts incident to underground coal mining. 30 C.F.R. 715.20 and 717.20 outline similar requirements for surface and underground coal mining in states without primacy, prior to OSM approval of a state regulatory program at least as stringent as the federal regulations.

[166] OSM, Directive 931, *Reforestation of Title IV and Title V Mined Lands*, June 10, 2008.

[167] FAA, "Wildlife and Habitat," http://www.environment.fhwa.dot.gov/ecosystems/wildlife/inv2dot.asp.

[168] See, for example, information at https://www.cbd.int/invasive/collaboration.shtml?org=icao.

[169] FHWA, "Plants and Invasive Species," http://environment.fhwa.dot.gov/ecosystems/vegmgmt_row.asp. Also see FHWA's *Guide for Local Highway and Street Maintenance Personnel*, August 2008, http://safety.fhwa.dot.gov/ local_rural/training/fhwasa07018/.

[170] NISC, U.S. Department of Treasury, http://www.invasivespecies.gov/main_nav/treasury.html.

[171] Other members of this committee include FWS, NOAA, EPA, the Department of Transportation Army Corps of Engineers, Coast Guard, and U.S. Geological Survey, as well as the Great Lakes states and the Great Lakes Fishery Commission, the Metropolitan Water Reclamation District of Greater Chicago, and the City of Chicago. For more information on this Committee, see http://asiancarp.us/.

[172] For more information, see http://www.whitehouse.gov/administration/eop/ostp/nstc/committees/cenrs.

[173] See CRS Report RL33472, *Sanitary and Phytosanitary (SPS) Concerns in Agricultural Trade.*

[174] NISC, "U.S. Agency for International Development," http://www.invasivespecies.gov/main_nav/usaid.html.

[175] §312(n) of the Clean Water Act. Regulations are at 40 CFR Part 1700. For more information, see EPA's website (http://water.epa.gov/lawsregs/lawsguidance/cwa/vessel/unds/FAQs.cfm).

[176] See, for example, NSF Press Release 09-173, "Invasive Species on the March: Variable Rates of Spread Set Current Limits to Predictability," September 17, 2009, http://www.nsf.gov/news/news_summ.jsp?cntn_id=115597.

[177] For more information, see SERC's website: http://www.serc.si.edu/labs/marine_invasions/index.aspx.

[178] For more information, see NBIC's website: http://invasions.si.edu/nbic/. http://www.serc.si.edu/labs/marine_invasions/index.aspx.

[178] For more information, see NBIC's website: http://invasions.si.edu/nbic/.

In: Invasive Species Management
Editor: Jeannie Saunders

ISBN: 978-1-63485-064-3

Chapter 3

INVASIVE SPECIES DEFINITION CLARIFICATION AND GUIDANCE WHITE PAPER*

Definitions Subcommittee

Preamble: Executive Order 13112 – defines an *invasive species* as "an alien species whose introduction does or is likely to cause economic or environmental harm or harm to human health." In the Executive Summary of the National Invasive Species Management Plan (NISMP) the term *invasive species* is further clarified and defined as "a species that is non-native to the ecosystem under consideration and whose introduction causes or is likely to cause economic or environmental harm or harm to human health." To provide guidance for the development and implementation of the NISMP, the National Invasive Species Council (NISC) and the Invasive Species Advisory Committee (ISAC) adopted a set of principles outlined in Appendix 6 of the NISMP. Guiding Principle #1 provides additional context for defining the term *invasive species* and states "many alien species are non-invasive and support human livelihoods or a preferred quality of life." However, some alien species (non-native will be used in this white paper because it is more descriptive than alien), for example West Nile virus, are considered invasive and undesirable by virtually everyone. Other non-native species are not as easily characterized. For example, some nonnative species are considered harmful, and therefore,

* This is an edited, reformatted and augmented version of a white paper issued by the Invasive Species Advisory Committee, as approved April 27, 2006.

invasive by some sectors of our society while others consider them beneficial. This discontinuity is reflective of the different value systems operating in our free society, and contributes to the complexity of defining the term *invasive species.*

NISC is engaged in evaluating and updating the 2001 NISMP and is developing comments for a revised action plan as required by the EO 13112. While there have been numerous attempts to clarify the term *invasive species,* there continues to be uncertainty concerning the use and perceived meaning of the term, and consequently over the prospective scope of actions proposed in the NISMP. Options related to private property use, pet ownership, agriculture, horticulture, and aquaculture enterprises may be affected depending upon the definition, use, and policy implications of the term.

Weeds as Examples

Weeds provide good examples to clarify what is meant by an invasive species because most people have a concept of what constitutes a "weed."

Invasion can be thought of as a process that in our example, a plant must go through to become a successful, yet harmful invader. Several barriers must be overcome for a plant to be considered an invasive weed. Invasive weeds are invasive species.

Large-scale geographical barriers

First, a geographical barrier must be overcome, which often occurs as a mountain range, ocean, or similar physical barrier to movement of seeds and other reproductive plant parts. Plants that overcome geographical barriers are known as alien plants or alien species. Alien plants are non-native plants and alien species are nonnative species. Therefore, non-native plants are those that occur outside their natural range boundaries, and this most often is mediated by humans either deliberately or unintentionally.

Survival barriers

The second set of obstacles that a non-native plant must overcome is barriers to germination and survival in its new location. These typically are environmental barriers such as adequate moisture availability to allow successful germination and survival of seedlings that will continue to grow to maturity. Other physical barriers might be soil pH, nutrient availability, or competition for resources from neighboring plants.

In particular, the desire to consider a non-native species as 'invasive' may trigger a risk/benefit assessment process to determine whether regulatory action is warranted. All these uncertainties have stood and could continue to stand in the way of progress in actions and policy development to prevent new invasions and manage existing invasive species. While it is not the purpose of this white paper to define a risk/benefit assessment process, development of such a process must be open and efficient to minimize the uncertainties.

This white paper is intended to provide a non-regulatory policy interpretation of the term *invasive species* by identifying what is meant, and just as important, what is not meant by the term. ISAC recognizes that biological and ecological definitions will not precisely apply to regulatory definitions. We believe, however, that our clarification will apply to *all taxa of invasive species in all habitats* and furthermore, our explanation will be functional and acceptable to most stakeholders. ISAC simply wants to clarify what is meant and what is not meant by the term invasive species in the technical sense and to provide insight into those areas where societal judgments will be necessary to implement effective public policy.

The utility of our clarification should be in education, conflict resolution, and efficiency in the planning, prevention, control/eradication, and management of *invasive species.*

ISAC recommends that NISC adopt the clarifications presented in this white paper to foster progress for invasive species management in the United States.

Weeds as Examples (continued)

Establishment barriers

The third obstacle that a non-native plant must overcome to be considered an invasive weed, is to form a population that is self-sustaining and does not need re-introduction to maintain a population base such that it continues to survive and thrive in its new environment. Once this occurs, this population of non-native plants is considered to be established. Environmental barriers to survival and establishment are similar.

Dispersal and spread barriers

Established non-native plants must overcome barriers to dispersal and spread from their site of establishment to be considered invasive plants. Additionally, the rate of spread must be relatively fast. However, this movement or spread alone does not necessarily make this non-native plant an invasive weed or invasive species.

Harm and impact

Finally, a plant is deemed to be invasive if it causes negative environmental, economic, or human health effects, which outweigh any beneficial effects. For example, yellow starthistle is a source of nectar for bee producers. But the displacement of native and other desirable plant species caused by yellow starthistle leads to dramatically decreased forage for wildlife and livestock, which severely disrupts the profitability of associated businesses. These negative effects greatly overshadow the positive effects and thus, define harm caused by yellow starthistle and explain why it is considered an invasive species.

INTRODUCTION

An invasive species is a non-native species whose introduction does or is likely to cause economic or environmental harm or harm to human, animal, or plant health. The National Invasive Species Management Plan indicates that NISC will focus on non-native organisms known to cause or likely to cause negative impacts and that do not provide an equivalent or greater benefit to society. In the technical sense, the term ‘invasion’ simply denotes the uncontrolled or unintended spread of an organism outside its native range with no specific reference about the environmental or economic consequences of such spread or their relationships to possible societal benefits. However, the policy context and subsequent management decisions necessitate narrowing what is meant and what is not meant by the term *invasive species.* Essentially, we are clarifying what is meant and not meant by “causing harm” by comparing negative effects caused by a non-native organism to its potential societal benefits.

PERCEPTION TO CAUSE HARM

Complications concerning the concept of *invasive species* arise from differing human values and perspectives. Differing perceptions of the relative harm caused or benefit gained by a particular organism are influenced by different values and management goals. If *invasive species* did not cause harm, we would not be nearly as concerned. Perceptions of relative benefit and harm also may change as new knowledge is acquired, or as human values or management goals change.

For a non-native organism to be considered an *invasive species* in the policy context, the negative effects that the organism causes or is likely to cause are deemed to outweigh any beneficial effects. Many non-native introductions provide benefits to society and even among species that technically meet the definition of invasive, societal benefits may greatly exceed any negative effects (for example crops and livestock raised for food). However, in some cases any positive effects are clearly overshadowed by negative effects, and this is the concept of causing harm. For example, water hyacinth has been popular in outdoor aquatic gardens but its escape to natural areas where its populations have expanded to completely cover lakes and rivers has devastated water bodies and the life they support, especially in the southeastern U.S. And, there are some organisms, such as West Nile virus, that provide almost no benefits to society at all. Such organisms constitute a small fraction of non-native species, but as a consequence of their ability to spread and establish populations outside their native ranges, they can be disastrous for the natural environment, the economies it supports, and/or public health. Because *invasive species* management is difficult and often very expensive, these worst offenders are the most obvious and best targets for policy attention and management.

The negative impact to a native species caused by an *invasive species* might trigger additional negative interactions for other associated native species; i.e., there could be direct and indirect effects. For example, an invasive weed that is undesirable as a food source may outcompete and displace native grasses and broadleaf plants. These displaced native grasses and broadleaf plants may have been primary forage for animals, which subsequently would be displaced to a new location or have their populations reduced because the weed invasion decreased the availability of food in their native plant and animal community. However, negative effects are not always characterized by a cascade of impacts realized throughout the environment. For example, simple displacement of an endangered species by a non-native species might alone provide sufficient justification to consider the non-native organism an *invasive species*.

WHAT WE DO NOT MEAN, WHAT WE DO MEAN, AND THE "GRAY" AREA

Native and Non-Native Species

Invasive species are species not native to the ecosystem being considered. Canada geese are native to North America and most of their populations migrate annually. However, in some locations in the U.S. (e.g. suburban Maryland; the Front Range of Colorado) introduced, non-migratory populations of Canada Geese are causing problems – such as fouling lawns, sidewalks, grass parks, and similar areas. While non-migratory populations can cause problems, they are not considered an invasive species because they are native. Additionally, Canada geese are of significant financial value to many local economies through waterfowl hunting and simple enjoyment. Mute swans, however, are invasive. Mute swans are native to Europe and Asia but were introduced into North America where their populations have increased dramatically. They compete directly with native waterfowl for habitat, displacing them, and that is why they are considered an invasive species. Whitetail deer populations have increased dramatically in the northeastern U.S. and are problems in farms, yards, and natural areas because they consume plants valued by humans; but are not invasive because they are native. Nutria, on the other hand, are another classic example of an invasive species. Nutria are native to South America but were introduced into North America where their populations have soared. Nutria compete directly with native muskrats, beavers, and other similar native species for habitat; often causing the displacement of these native species.

Feral Populations

It is also essential to recognize that *invasive species* are not those under human control or domestication; that is, *invasive species* are not those that humans depend upon for economic security, maintaining a desirable quality of life, or survival. However, the essential test is that populations of these species must be under control. Escaped or feral populations of formerly domesticated plants and animals would be considered *invasive species* if all the concepts and conditions are met as outlined in "Weeds Are Examples." Cereal rye being produced on a farm in Kansas is considered very desirable, but feral rye on the breaks of the Poudre River in Colorado would be considered an *invasive species* because it is displacing native plants and the native animal communities they support. Domesticated goats on a farm in Texas are considered highly desirable, but feral goats in Haleakala National Park on Maui are considered an *invasive species.* Feral goats have severely overgrazed areas and eliminated native Hawaiian plants, which were never adapted to grazing. Areas denuded by feral goats have led to increased soil erosion.

A Biogeographical Context

An *invasive species* may be invasive in one part of the country, but not in another. A biogeographical context must be included when assessing whether a non-native species should be considered an *invasive species.* Lake trout are highly desirable in the Great Lakes where they are native, but are considered an *invasive species* in Yellowstone Lake. They compete with native cutthroat trout for habitat, which decreases their populations. Atlantic saltmarsh cordgrass is an essential component of east coast salt marshes, but is highly invasive on the west coast where it covers mudflats and displaces native estuarine plants and the community of animals they support, including huge flocks of migrating waterfowl. Kentucky bluegrass would be considered an *invasive species* in Rocky Mountain National Park in Colorado, but considered non-invasive a mere 60 miles away at a golf course in Denver. English ivy is considered a good ground cover species in the Great Plains and Midwest, but is a highly invasive weed in the forests of the Pacific Northwest and Eastern U.S. where it outcompetes native plants and displaces the associated animal communities.

The "Gray" Area

There are obvious examples of *invasive species* such as snakehead fish, yellow starthistle, or *Phytophthora ramorum* (the organism that causes sudden oak death); and there are obvious examples of species that are not invasive, namely native plants and animals. There are, however, non-native organisms for which it will be difficult to make a determination and these should be subject to assessment. Whether these non-native organisms will be considered *invasive species* will depend upon human values. For example, European honeybees are cultured to produce honey and pollination services, and even though they form wild populations in many parts of the country and occasionally create problems by building hives in the walls of homes or can be a human health problem for individuals that are highly allergic to their sting, most would not consider them an *invasive species* because they produce a desired food product.

Another gray area example would be native termites v. Formosan termites. No one wants termites in their homes but only Formosan termites would be considered an *invasive species* because they are non-native. Smooth brome also serves as another gray area example. It was imported from Russia in the 1890s for forage and was widely planted. It clearly has escaped cultivation and can be found in many natural areas particularly in the western U.S. but in most situations, smooth brome would not be considered an *invasive species* because of its forage value for wildlife and livestock.

Chinese or Oriental clematis serves as another gray area example. Chinese clematis (virgin's bower, orange peel) is a popular ornamental that has been planted worldwide. However, it has escaped cultivation in several western states where its populations can spread, particularly in shrubland, on riverbanks, sand depressions, along roadsides, in gullies, and along riparian forests in hot dry valleys, deserts, and semi-desert areas. Escaped populations of Chinese clematis occur in Idaho, Nevada, Utah, New Mexico, and Colorado but so far, it is considered an *invasive species* only in Colorado where it has spread dramatically from its site of introduction and displaced native plant species.

Environmental Harm

We use environmental harm to mean biologically significant decreases in native species populations, alterations to plant and animal communities or to ecological processes that native species and other desirable plants and animals and humans depend on for survival. Environmental harm may be a result of direct effects of *invasive species*, leading to biologically significant decreases in native species populations.

Examples of direct effects on native species include preying and feeding on them, causing or vectoring diseases, preventing them from reproducing or killing their young, out-competing them for food, nutrients, light, nest sites or other vital resources, or hybridizing with them so frequently that within a few generations, few if any truly native individuals remain. Environmental harm includes decreases in populations of Federally Listed Threatened and Endangered Species, other rare or uncommon species and even in populations of otherwise common native species. For example, over three billion individual American chestnut trees were found in U.S. forests before the invasive chestnut blight arrived and virtually eliminated them. Environmental harm also can be the result of an indirect effect of invasive species, such as the decreases in native waterfowl populations that may result when an invasive wetland plant decreases the abundance of native plants and thus, decreases seeds and other food that they provide and that the waterfowl depend upon.

Environmental harm also includes significant changes in ecological processes, sometimes across entire regions, which result in conditions that native species and even entire plant and animal communities cannot tolerate. For example, some non-native plants can change the frequency and intensity of wildfires, or alter the hydrology of rivers, streams, lakes and wetlands and that is why they are considered *invasive species*. Others can significantly alter erosion rates. For example, trapping far more wind-blown sand than native dune species, or holding far less soil than native grassland species following rainstorms. Some invasive plants and micro-organisms can alter soil chemistry across large areas, significantly altering soil pH or soil nutrient availability. Environmental harm also includes significant changes in the composition and even the structure of native plant and animal communities. For example, the invasive tree *Melaleuca quinquinervia*, can spread into and take over marshes in Florida's Everglades, changing them from open grassy marshes to closed canopy swamp-forests.

Environmental harm may also cause or be associated with economic losses and damage to human, plant and animal health. For example invasions by fire promoting grasses that alter entire plant and animal communities eliminating or sharply reducing populations of many native plant and animal species, can also lead to large increases in fire-fighting costs and sharp decreases in forage for livestock. West Nile virus is a well known human health problem caused by a non-native virus which is commonly carried by mosquitoes. West Nile Virus also kills many native bird species, causing drastic reduction in populations for some species including crows and jays.

ADDITIONAL EXAMPLES OF IMPACTS CAUSED BY INVASIVE SPECIES

Specific examples of the harm caused by *invasive species* are useful to further clarify the definition. The following list of examples is not meant to be comprehensive, but offers further explanation:

IMPACTS TO HUMAN HEALTH

Respiratory infections: The outbreak of *West Nile virus* in the U.S. began in the Northeast in 1999 and has since spread throughout the country. Infections in humans may result in a flu-like illness and in some cases death. This outbreak has caused illness in thousands of citizens, increased medical costs for affected persons, and decreased productivity due to absence from work. *West Nile virus* also has affected horses and has caused widespread mortality in native birds (U.S. Centers for Disease Control, 2006).

Poisonous plants: Exposure to the sap of *Tree-of-heaven/Chinese sumac tree* has caused inflammation of the heart muscle (myocarditis) in workers charged to clear infested areas. Afflicted personnel experienced fever/chills, chest pain that radiated down both arms, and shortness of breath. Exposure occurred when sap from tree-of-heaven contacted broken skin. Such exposure has caused hospitalization, medical expense, and lost productivity due to absence from work (Bisognano et al. 2005).

IMPACTS TO NATURAL RESOURCES

Declines in wildlife habitat and timber availability: *Chestnut blight* is a disease of American chestnut caused by a non-native fungal pathogen that was introduced into eastern North America around 1910. The disease eliminated the American chestnut from eastern deciduous forests thereby decreasing timber harvests and wildlife that depended upon the American chestnut for habitat (USDAAPHIS/FS 2000).

European gypsy moth defoliates trees on millions of acres of northeastern and mid-western forests. It currently is found in 19 states causing an estimated \$3.9 billion in tree losses and also decreased wildlife habitat (USDA-APHIS/FS 2000).

Decreased soil stabilization and interrupted forest succession: *White pine blister rust* is a disease of white pine species caused by the non-native fungal pathogen *Cronartium ribicola*. It was introduced into eastern North America around 1900 and western North America in 1920. It spread rapidly, killing off native white, whitebark, and limber pines, whose seeds are an important food source for birds, rodents and bears. Elimination of these trees caused by this pathogen alters forest ecosystems, eliminates wildlife forage, and decreases the soil stabilization effects of these trees, snowmelt regulation, and forest succession (Krakowski et al. 2003).

Changes in wildfire frequency and intensity: *Cheatgrass* decreases the interval between the occurrences of wildfires in the Great Basin region from once every 70 to 100 years to every 3 to 5 years because it forms dense stands of fine fuel annually. The decrease in interval between wildlfires causes increased risk to human life and property and also places at risk established communities of plants and animals that we consider desirable (Knapp 1996; Pimentel et al. 2000; USFWS 2003; Whisenant 1990).

Excessive use of resources: *Tamarisk* in the desert southwest use more than twice as much water annually as all the cities in southern California, which places this invasive weed in direct competition with humans for the most limiting resource in the southwestern U.S. (Friederici 1995; Johnson 1986).

Suppressors: Russian knapweed exudes toxins from its tissues that inhibit the growth of surrounding plants or eliminates them. Desirable plant communities are placed at risk from *Russian knapweed* invasion, which may result in decreased numbers of wildlife species or livestock that the invaded land otherwise could support. *Russian knapweed* also is very toxic to horses (Stevens 1986; Young et al. 1970a and 1970b).

Decreased carrying capacity for wildlife and livestock: Expansion of *leafy spurge*, *yellow starthistle*, or other unpalatable invasive weeds displace desirable forage plants and may allow fewer grazing animals to survive in infested areas (DiTomaso 2001; Lym and Messersmith 1985; Lym and Kirby 1987).

IMPACTS TO RECREATIONAL OPPORTUNITIES AND OTHER HUMAN VALUES

Decreased property values: Asian longhorned beetles first appeared in New York in 1996 and in Chicago in 1998. Larvae burrow into trees causing girdling of stems and branches, dieback of the crown, and can kill an entire tree. It infests many different tree species in the U.S. and is a threat to urban and rural forests (Cavey et al. 1998).

Emerald ash borers were first detected in the U.S. in 2002. They currently are found in Michigan, Ohio, and Indiana. Emerald ash borer larvae tunnel under bark of ash trees and could eliminate ash as a street, shade, and forest tree throughout the U.S. Estimated replacement cost in six Michigan counties is $11 billion and an additional $2 million in lost nursery sales (Chornesky et al. 2005).

Dutch elm disease was first introduced into the U.S. in 1927 and occurs in most states. Dutch elm disease has killed more than 60% of elms in urban settings and decreased the value of urban and suburban properties (Brasier and Buck 2001).

Spotted knapweed and *leafy spurge* expansion in the western U.S. have displaced desirable forage plants thereby decreasing the value and sales price of grazingland in the western U.S. (Maddox 1979; Weiser 1998).

Eurasian watermilfoil was introduced into the U.S in the 1940s and has since spread throughout much of the country. This submersed aquatic plant can form dense mats at the water surface limiting access, recreation, and aesthetics and thus, has decreased the values of shoreline properties in New Hampshire, the Midwest and elsewhere (Halstead et al. 2003).

Decreased sport fishing opportunities: Whirling disease is caused by a parasite (*Myxobolus cerebralis*) that most likely originated in Europe. It was first observed in the U.S. in 1958. The parasite attacks the soft cartilage of young trout causing spinal deformities and causes the fish to exhibit erratic tail-chasing behavior. Heavily infected young trout can die from *Whirling disease* and even if they recover, they remain carriers of the parasite. All species of trout and salmon may be susceptible and angling and the businesses supported by trout and salmon fishing may be at risk if this disease continues to spread (Aquatic Nuisance Species Task Force et al. 2005; Colorado Division of Wildlife 2006).

Smallmouth bass fishing in Lake Erie was closed during bass mating because of *round goby* predation of nests. Fishing was closed because male smallmouth bass aggressively guard nests from predators and are easier to catch by anglers during this time of year. Removal of males by anglers decreased the number of bass offspring because of increased *round goby* predation of unguarded nests (Steinhart et al. 2004). Businesses that smallmouth bass anglers patronize could be adversely affected by such closures.

Altered business opportunities: The concern over *Sudden Oak Death Syndrome* caused by the pathogen *Phytophthora ramorum* is causing drastic changes in available nursery stock by nurseries and landscape businesses. This clearly impacts the profitability of these businesses and choice by consumers and could devastate oak forests nationwide (Chornesky et al. 2005; Rizzo and Garbelotto 2003).

Annual harvests of oysters in Long Island Sound averaged over 680,000 bushels during 1991 through 1996. After *Haplosporidium nelsonii* (MSX) invaded in 1997 and 1998, oyster harvests decreased from 1997 through 2002 to an average annual harvest of 119,000 bushels with a low of 32,000 bushels in 2002. The overall ex-vessel value of oyster farming dropped 96% in 10 years from $45 million in 1992 to $2 million in 2002 (Sunila et al. 1999).

Non-native algae introduced into the Hawaiian Islands costs Maui alone about $20,000,000 annually due to algae fouling the beaches and subsequent lost tourism (Carroll 2004; Keeney 2004; Univ. Hawaii 2006).

Sea lampreys were introduced into Lakes Ontario and Erie during the construction of the Welland Canal and quickly spread to the other Great Lakes. The *sea lamprey* is a parasite that attaches itself to fish, eventually killing them, and has devastated commercial and recreational Lake Trout fishing in the Great Lakes (Lawrie 1970).

Australian spotted jellyfish were introduced into the Gulf of Mexico in 2000 and occurred in such massive numbers that shrimping operations were shut down because jellyfish clogged shrimp nets (Graham et al. 2003).

Altered ecosystems and recreational opportunities: The submersed aquatic plant *hydrilla,* forms dense canopies at the water surface that raise surface water temperatures, change pH, exclude light, and consume oxygen, resulting in native plant displacement and stunted sport fish populations. This example of an altered aquatic ecosystem caused by an invasive aquatic weed also negatively affects recreation and businesses that depend upon that human activity (Colle et al. 1987).

CONCLUSION

Invasive species are those that are not native to the ecosystem under consideration and that cause or are likely to cause economic or environmental harm or harm to human, animal, or plant health. Plant and animal species under domestication or cultivation and under human control are not *invasive species*. Furthermore for policy purposes, to be considered invasive, the negative impacts caused by a non-native species will be deemed to outweigh the beneficial effects it provides. Finally, a non-native species might be considered invasive in one region but not in another. Whether or not a species is considered an *invasive species* depends largely on human values. By attempting to manage *invasive species*, we are affirming our economic and environmental values. Those non-native species judged to cause overall economic or environmental harm or harm to human health may be considered invasive, even if they yield some beneficial effects. Society struggles to determine the appropriate course of action in such cases, but in a democratic society that struggle is essential.

Many *invasive species* are examples of "the tragedy of the commons," or how actions that benefit one individual's use of resources may negatively impact others and result in a significant overall increase in damage to the economy, the environment, or public health. In ISAC's review of Executive Order 113112, the public domain is specifically represented; however, the implementation of the NISMP has prompted concerns over the rights of personal and private property owners. Property rights are of great importance in the U.S. and one outcome of the NISMP should be to recognize the right to self determination by property owners and promote collaboration on *invasive species* management. The right to self determination is an important concept in a democratic society, however, with that right comes personal responsibility and stewardship, which includes being environmentally responsible. The natural environment that our society enjoys, recreates in, and depends upon to support commerce must be conserved and maintained. Effective *invasive species* management is just one aspect of conserving and maintaining our nation's natural environment, the economies it supports, and the high quality of life our society enjoys.

REFERENCES

Aquatic Nuisance Species Task Force, U.S. Fish and Wildlife Service, and U.S. Coast Guard. 2005. Protect Your Waters. Harmful Aquatic Hitchikers: Others: Whirling Disease. [Online] http://protectyour waters.net/hitchhikers/others_whirling_disease.php.

Bisognano, J.D., K.S. McGrody, and A.M. Spence. 2005. Myocarditis from the Chinese sumac tree. Annals Internal Medicine 143(2):159.

Brasier, C.M. and K.W. Buck. 2001. Rapid evolutionary changes in a globally invading fungal pathogen (Dutch elm disease). *Biological Invasions* 3:223-233.

Carroll, R. 2004. Maui battling seaweed invasion. Assoc. Press. [Online] http://thehonoluluadvertiser.com/artucke/jan/05/ln/ln10a.htm.

Cavey, J.F., E. R. Hoebeke, S. Passoa, and S.W. Lingafelter. 1998. A new exotic threat to North American hardwood forests: an asian longhorned beetle, Anoplophora glabripennis (Motschulsky) (Coleoptera: Cerambycidae). I. Larval description and diagnosis. *Proc. Entomol. Soc. Wash.* 100 (2):373-381.

Chorensky, E.A., A.M. Bartuska, G.H. Aplet, K.O. Britton, J. Cummings-Carlson, F.W. Davis, J. Eskow, D.R. Gordon, K.W. Gottschalk, R.A. Haack, A.J. Hansen, R.N. Mack, R.J. Rahel, M.A. Shannon, L.A. Wainger, and T.B. Wigley. 2005. Science priorities for reducing the threat of invasive species to sustainable forestry. *BioSci.* 55(4):335-348.

Colle, D.E., J.V. Shireman, W.T. Haller, J.C. Joyce, and D.E. Canfield. 1987. Influence of Hydrilla on Harvestable Sport-Fish Populations, Angler Use, and Angler Expenditures at Orange Lake, Florida. North American *Journal of Fisheries Management* 7:410-417.

Colorado Division of Wildlife. 2006. Whirling disease and Colorado's trout. [Online] http://wildlife.state.co.us/fishing/whirling.asp.

DiTomaso, J. 2001. Element stewardship abstract: Centaurea solstitialis L. Weeds on the web: The Nature Consevancy wildland invasive species program. [Onlin] http://tncweeds.ucdavis.edu/esadocs/docmnts/centsols. Htm.

Friedercici, P. 1995. The alien saltcedar. Am. For. 101:45-47.

Graham, W.M., D.L. Martin, D.L. Fedder, V.L Asper, and H.M. Perry. 2003. Ecological and economic implications of a tropical jellyfish invader in the Gulf of Mexico. *Biological Invasions* 5(1-2) 53-69.

Halstead, J.M., J. Michaud, S. Hallas-Burt, and J.P Gibbs. 2003. Hedonic analysis of effects of a nonnative invader (Myriophyllum heterophyllum) on New Hampshire (USA) lakefront properties. *Environmental Management* 32(3): 391-398.

Johnson, S. 1986. Alien plants drain western waters. The Nature Conservancy News, Oct-Nov 1986.

Keeney, T.R.E. 2004. Written testimony before the Subcommittee on Fisheries Conservation, Wildlife and Oceans, Committee on Resources, U.S. House of Representatives. http://www.ogc.doc.gov/ogc/legreg/testimon/108s/keeney0415.htm.

Knapp, P.A. 1996. Cheatgrass (Bromus tectorum L.) dominance in the Great Basin Desert: History, persistence, and influences to human activities. Global Environ. Change 6(1):37-52.

Krakowski, J., S.N. Aitken, Y.A. El-Kassaby. 2003. Inbreeding and conservation genetics in whitebark pine. *Conservation Genetics* 4:581-593.

Lawrie, A.H. 1970. The sea lamprey in the Great Lakes. *Transactions of the American Fisheries Society* 99:766-775.

Lym, R.G. and C.G. Messersmith. 1985. Cost effectiveness of leafy spurge control during a five-year management program. *North Dakota Farm Res.* 43(1)7-10.

Lym, R.G. and D.R. Kirby. 1987. Cattle foraging behavior in leafy spurge infested rangeland. *Weed Technol.* 1:314-318.

Maddox, D.M. 1979. The knapweeds: Their economics an biological control in the western states, U.S.A. Rangelands 1(4):139-141.

Pimentel, D., L. Lach, R. Zuniga, and D. Morrison. 2000. Environmental and economic costs of non-indigenous species in the United States. *Biosci.* 50(1):53-65.

Richardson, D.M., P. Pysek, M. Rejmanek, M.G. Barbour. F.D. Panetta, and C.J. West. 2000.

Naturalization and invasion of alien plants: concepts and definitions. *Diversity and Distributions* 6:93- 107.

Rizzo, D.M. and M. Garbelotto. 2003. Sudden oak death: Endangering California and Oregon forest ecosystems. *Frontiers in Ecology and the Environment* 1:197-204.

Steinhart, G.B., E.A. Marschall, and R.A. Stein. 2004. Round goby predation on smallmouth bass offspring in nests during simulated catch-and-release angling. *Transactions of the American Fisheries Society* 133: 121-131.

Stevens, K.L. 1986. Alleopathic polyacetylenes from Centaurea repens (Russian knapweed). *J. Chem. Ecol.* 12:1205-1211.

Sunila, I., J. Karolus, and J. Volk. 1999. A new epizootic of Haplosporidium nelsoni (MSX), a haplosporidian oyster parasite, in Long Island Sound, Connecticut. *Journal of Shellfish Research* 18(1): 169-174.

U.S. Fish and Wildlife Service, 2003. Costly fires hurt wildlife habitat. [Online] http://www.fws.gov/deerflat/currfire.htm.

Univ. of Hawaii. 2006. Invasive marine algae of Hawaii. [Online] http://www.hawaii.edu/reefalgae/invasive_algae/INDEX.HTM.

U.S. Center for Disease Control. 2000. Fight the Bite. [Online] http://www.cdc.gov/nicdod/dvbid/westnile/index.htm.

USDA-APHIS/FS. 2000. Draft pest risk assessment for importation of solid wood packing materials into the United States. http://www.aphis.usda.gov/ppq/pra/swpm/.

Weiser, C. 1998. Economic effects of invasive weeds on land values (From and agricultural banker's standpoint). Proc. Colorado Weed Summit April 7-8, 1998. p. 35-38.

Whisenant, S.G. 1990. Changing fire frequencies on Idaho's Snake River Plain: Ecological and management implications. The Station. Nov. 1990 (276) Ogden, UT: General Technical Report INT – USDA Forest Service Intermountain Research Station.

Young, S, W.W. Brown, and B. Klinger. 1970a. Nigropallidal encephalomalacia in horses caused by ingestion of weeds of the genus Centaurea. *J. Amer. Vet. Med. Assoc*. 157:1602-1605.

Young, S., W.W. Brown, and B. Klinger. 1970b. Nigropallidal encephalomalacia in horses fed Russian knapweed (Centaurea repens L.). Amer. *J. Vet. Res.* 31:1393-1404.

In: Invasive Species Management
Editor: Jeannie Saunders
ISBN: 978-1-63485-064-3

Chapter 4

STATEMENT OF PAUL REIS, ASSOCIATE DEPUTY CHIEF, UNITED STATES FOREST SERVICE. HEARING ON "INVASIVE SPECIES MANAGEMENT ON FEDERAL LANDS"*

Mr. Chairman and Members of the Subcommittee, thank you for the opportunity to testify before you today on the role of the Forest Service in protecting forests and grasslands from invasive species. The Forest Service is committed to the prevention, detection, control, management and eradication of invasive species, and to restoring the structure and function of affected aquatic and terrestrial ecosystems on all lands.

BACKGROUND

Invasive species are among the most significant environmental and economic threats facing our Nation. Aquatic and terrestrial invasive plants, pathogens, vertebrates, invertebrates, algae, and fungi have become established on millions of acres across North America. These infestations are degrading watershed condition and ecosystem functionality, reducing forest and rangeland productivity, increasing the risk of wildfire and soil erosion, causing declines in recreational use and enjoyment, negatively impacting

* This is an edited, reformatted and augmented version of a statement presented May 16, 2013 before the House Committee on Natural Resources, Subcommittee on Public Lands and Environmental Regulation.

human health and safety, threatening native fish and wildlife populations and their associated habitats, causing declines in property values, and undermining the economy at all levels. Invasive species cause billions of dollars in damage each year in the United States. Pimentel et al. (2001) estimated damage from invasive species worldwide totaled at more than $1.4 trillion per year.

Burgeoning global trade and transportation have facilitated the spread of many species among continents well beyond their native range. With the number of people living in, enjoying, and using forests, grasslands, and water resources continually increasing, the likelihood of invasive species spreading through transportation and recreational activities is also rising. As a result, many species of invasive plants, pathogens, vertebrates, invertebrates, and other harmful exotic species have been introduced to our Nation's aquatic and terrestrial ecosystems. Many of these have become established within these ecosystems.

Responsibilities and Capabilities of the Forest Service

The Forest Service plays an important role in the Nation's efforts to address the threat of invasive species across the landscape through our National Forest System, State and Private Forestry, Research and Development, and International program areas. In this testimony we will explore how individually and collectively these programs work together to address invasive species threats.

With internationally recognized land management and scientific expertise, the Forest Service is well suited to address the many challenges of invasive species. The Forest Service continues to play an important national and international leadership role in advancing the understanding of the invasive species problem. The wide ranging authorities of the Forest Service allow us to work with partners to combat invasive species across all lands, public and private. We also develop methods, tools, and approaches, through which these harmful exotic species can be detected, prevented, controlled, and eradicated.

At the national, regional, state and local levels the Forest Service works extensively with county, state, tribal, federal, and private stakeholders to proactively implement invasive species management activities across the broader landscape. Through an "all lands approach" the Forest Service provides a wide range of technical and financial assistance to help manage invasive species. The Forest Service works closely with State forestry agencies to implement State Forest Action Plans to protect forest from threats.

grasslands, (3) conduct research, and (4) provide technical assistance to the private sector and in large agricultural pest control projects.

The U.S. Department of Agriculture (USDA), Agricultural Research Service (ARS) conducts research in extremely diverse areas involving prevention, control and management of invasive species. For example, ARS provides research in support of action agencies such as the Animal and Plant Health Inspection Service (APHIS), to reduce the rate of introduction of invasive species, and to rapidly detect, identify and eradicate incipient species.

The Animal and Plant Health Inspection Service (APHIS) is a multi-faceted Agency with a broad mission area that includes protecting and promoting U.S. agricultural health, regulating genetically engineered organisms, administering the Animal Welfare Act and carrying out wildlife damage management activities. APHIS' mission has expanded over the years to include protection of public health and safety as well as natural resources that are vulnerable to invasive pests and pathogens.

The Natural Resource Conservation Service (NRCS) has become a conservation leader for all natural resources, ensuring private lands are conserved, restored, and more resilient to environmental challenge. NRCS helps private landowners tackle invasive species problems in four major ways: 1) technical and financial assistance to manage invasive species and pests; 2) conservation initiatives that work at a landscape scale to address natural resource concerns, including invasive species; 3) Conservation Innovation Grants with partner entities to support development and implementation of innovative approaches and strategies to address invasive species; and 4) Plant Materials Center research geared toward invasive species management and restoring areas where invasive species have been removed.

The Forest Service Forest Health Protection programs direct and implement measures to prevent, detect, contain, and suppress unwanted native and invasive insects, pathogens, and plants affecting trees and forests. In FY 2012, State and Private Forestry programs provided $1.4 million in essential matching funds and technical assistance to state governments to combat economically significant weed threats to state and private forest lands.

Forest Health Protection and partners from cooperating states conduct an annual collaborative forest pest survey on over 400 million acres of forest land. We recently completed high resolution maps for over 250 forest tree species in the US that will be used to guide these surveys in the future. Additionally, we have developed a program called the "Forest Disturbance Mapper," a near real-time web portal that uses remote sensing to detect disturbances caused by forest pests, and an interagency database to detect and track thousand cankers disease of walnut and other pests.

In FY 2012, Forest Service *Research and Development* delivered 169 invasive species tools including the identification of key pathways for invasion by new forest pests; methods for detecting, monitoring, and controlling the walnut twig beetle; release and recovery guidelines for biological control agents for emerald ash borer; and an assessment of the potential impacts of hemlock woolly adelgid predators.

The Forest Service *International Programs* also work to protect our forests from invasive species damage. For example, the program works with Chinese counterparts who have partnered with us to address one of the most destructive invasive forest pests, the emerald ash borer (EAB). The Forest Service continues to work with the USDA Agricultural Research Service (ARS) to better understand why the borer is so resilient and pervasive. This will help predict and prevent potential future outbreaks by related wood boring beetles. With an aim of identifying biocontrol mechanisms, a partnership was formed between the Forest Service's Northern Research Station, the ARS and counterparts in China. With support from International Programs, the team is working to find natural enemies of EAB in its native range.

Strategic Approach to Invasive Species Management

To ensure the continued production of needed goods, services, and values from our Nation's terrestrial and aquatic ecosystems, the Forest Service takes a strategic approach for managing invasive species across all program areas. This approach includes four main elements: *(1) prevention, (2) detection, (3) control and management, and (4) restoration and rehabilitation.*

Prevention

The most effective strategy to protect forests, waterways, and grasslands from invasive species is to prevent invasive species introduction and establishment. Containing known infestations is also important for blocking the spread of invasive species from infested lands to surrounding areas. We coordinate with Federal and State regulatory agencies to understand pathways for introductions, implement quarantine regulations, survey for invasive species, and educate the public about invasive pest threats and how to prevent the spread of invasive species.

Forest Service researchers in partnership with APHIS are working with industry partners to reduce the introduction of invasives into the US through shipments of wood products and packaging and the live plant trade. Additionally, Forest Service scientists and managers at the Eastern and Western Threat Centers are working closely with domestic and international partners to develop a comprehensive database for prediction, prevention, and proactive management of invasive plants. A public education campaign developed by the Forest Service in partnership with Wildlife Forever recruits hunters, anglers, and recreational boaters to help prevent the spread of aquatic invasive species such as quagga and zebra mussels and Eurasian milfoil.

Detection

The Forest Service develops and implements efficient survey and monitoring tools and technologies to facilitate early detection of invasive species, including in urban areas, and rapidly assess their potential impact on forest and grassland health. As necessary and appropriate, the Forest Service coordinates these activities with Federal and non-Federal cooperators across all lands.

The Forest Service has supported development of a mapping system used nationally by cooperating agencies and weed management organizations to document distribution of invasive weeds. Additionally, Forest Service scientists developed a test capable of detecting the fungal pathogen that causes white-nose syndrome (WNS) in bats. The test is being used to identify infested caves, so that Forest Service and other land managers might selectively restrict access to those caves and mines to help slow the spread of WNS.

Control and Management

The Forest Service directly intervenes to manage populations of invasive species that threaten forest and grassland health and sustainability. Rapid response following early detection is used to eradicate new infestations. If eradication is not feasible, Integrated Pest Management (IPM) and adaptive management techniques are implemented to help maintain ecosystem function. This includes research and management to increase the resilience of threatened ecosystems to mitigate the impacts of pests. In cooperation with external stakeholders, the Forest Service conducts research to characterize infestations, to identify factors conducive to infestations, and to develop tools and techniques to cost-effectively eradicate or manage priority invasive species.

For example, the Jackson and Buffalo Ranger Districts of the Bridger-Teton National Forest in Wyoming include the majority of the land within the Jackson Hole Weed Management Association, where the Forest Service identified approximately 7,000 priority acres for detection and immediate eradication efforts. In total, the Forest Service successfully eradicated 15 priority species from those 7,000 acres. Since 2000, the Forest Service, working in partnership with States and other Federal agencies, has implemented a national Slow the Spread (STS) strategy to minimize the rate at which gypsy moth spreads into uninfested areas. The STS program has reduced the spread of gypsy moth more than 60 percent from the historical level of 13 miles per year. In only 12 years, this program has prevented impacts on more than 100 million acres. When oak trees started dying in the San Francisco Bay Region, the Forest Service Pacific Southwest Research Station developed a collaborative research response that helped identify the cause—a water mold previously unknown to science. The combined efforts of the Forest Service with APHIS and numerous partners via the California Oak Mortality Task Force have reduced the human-assisted spread of Sudden Oak Death and helped communities in the 14 infested coastal counties in California and Oregon deal with the infestation.

Restoration and Rehabilitation

Restoring landscapes that have been impacted by invasive species or associated management activities is necessary for improving ecosystem integrity and function and may reduce vulnerability to invasive species establishment in the future. Restoring and maintaining the health, functions, and productivity of areas affected by invasive species is consistent with management guidance on restoring national forests and the effective use of native species.

For example, In order to restore cutthroat trout populations to streams, non-native trout are replaced with genetically pure cutthroat populations. After a decade of restoration efforts, Cherry Creek, on the Gallatin National Forest, now contains the largest genetically pure population of this cutthroat trout subspecies in the upper Missouri River drainage area.

CONCLUSION

In conclusion, the invasive species issue is considered a high priority by all program areas of the U.S. Forest Service. We believe the Forest Service collaborative approach to invasive species management enhances our ability to work together by building on each other's strengths and authorities. In addition, our Forest Service personnel works with local, county, and State governments; Cooperative Weed Management Areas; Cooperative Invasive Species Management Areas; our departmental partners NRCS, ARS and APHIS; and other organizations in the public and private sectors to promote a collaborative approach to mitigate, manage, and if necessary, adapt to invasive species threats across the landscape.

I would like to thank the committee members for their interest in invasive species management, and I welcome any questions you may have for me at this time.

In: Invasive Species Management
Editor: Jeannie Saunders
ISBN: 978-1-63485-064-3

Chapter 5

TESTIMONY OF RANDY DYE, PRESIDENT, NATIONAL ASSOCIATION OF STATE FORESTERS. HEARING ON "INVASIVE SPECIES MANAGEMENT ON FEDERAL LANDS"*

On behalf of the National Association of State Foresters, I thank Chairman Bishop and Ranking Member Grijalva for the opportunity to appear before the Subcommittee today. The National Association of State Foresters (NASF) represents the directors of the state forestry agencies of all fifty states, eight territories, and the District of Columbia. State Foresters manage and protect state and private forests across the U.S., which make up two-thirds of the nation's forests, and work closely with our federal partners to respond to invasive species issues.

The impacts of invasive forest insect and disease species on our nation's forests have become an increasing concern for the National Association of State Foresters (NASF). NASF's programs and stewardship actions are on the front lines of eradicating, slowing the spread, and addressing the enormous collateral damage of invasive species.

Forested landscapes cover approximately one-third of the total land area of the United States, including 100 million acres in urban environments. Every American benefits from forests, whether in the form of wood products for

* This is an edited, reformatted and augmented version of testimony presented May 16, 2013 before the House Committee on Natural Resources, Subcommittee on Public Lands and Environmental Regulation.

construction or paper, neighborhood amenities, wildlife habitat, carbon sequestration, clean water and air, and even our spiritual well-being. Many Americans' jobs are linked to trees. The U.S. forest products industry employs nearly 900,000 people; it is among the top 10 manufacturing sector employers in 47 states. Jobs associated with production of non-wood forest products are estimated to be in the tens of thousands.

Invasive species know no boundaries; they span landscapes, land ownerships, and jurisdictions. The damage they cause costs the American public an estimated $138 billion each year, which makes them a significant drain on the national economy.

- Private landowners and small communities are some of the hardest hit by invasive species infestations.
- Invasive species can be exceptionally damaging in urban environments where ecological systems are already stressed. Invasive species threaten the quality of life and the property values of millions of metropolitan residents across the country.
- Currently, 42 percent—400 of 958—of the plant and animal species listed by the Federal Government as threatened or endangered have been negatively affected by invasive species.
- Invasive species populations have depleted water supplies, poisoned wildlife and livestock, and directly impacted thousands of acres of native forests and rangelands.
- Public recreational opportunities and experiences have become severely degraded by rapid infestations of invasive species, in many cases hampering access, reducing recreational quality and enjoyment, and decreasing the aesthetic values of public lands

Some of the most damaging Invasive species include Asian Long-horned Beetle, Emerald Ash Borer, Gypsy Moth, Sudden Oak Death, Hemlock Woolly Adelgid, and Cogon grass. Municipal governments across the country are spending more than $1.7 billion each year to remove trees on city property killed by these pests. Homeowners are spending $1 billion to remove and replace trees on their property and they are absorbing an additional $1.5 billion in reduced property values. The scope of the impacts of these pests is demonstrated by a brief description of the threats they pose:

- The Asian Longhorned Beetle kills trees in 15 botanical families—especially maple and birch which constitute much of the forest reaching from Maine to Minnesota and urban trees worth an estimated $600 billion.

- Emerald Ash Borer occupies more than 200,000 square miles in 18 states. More than 200 million ash trees in the Plains States and additional trees in the South are at risk to this pest. Homeowners and municipalities collectively will pay more than $10 billion over the next 10 years to remove dead ash trees that would otherwise fall and could cause property damage or even loss of life.
- Hemlock Woolly Adelgid has killed up to 90% of hemlock trees in the Appalachians from Georgia to Massachusetts. Loss of hemlock groves threatens unique ecosystems and watersheds.
- Goldspotted Oak Borer has killed up to 80,000 California live oak and black oak trees in San Diego County in less than 15 years. The insect threatens oaks throughout California, including close to 300,000 oak trees growing in greater Los Angeles and Yosemite Valley.
- Sudden Oak Death affects 143 different plant species and continues to spread in California's 14 impacted counties as well as Curry County, Oregon. In 2012 alone, nearly 400,000 trees were lost to Sudden Oak Death in California.

The Federal government has several unique characteristics that compel it to play a primary role in the fight against invasive species. It has the direct authority to manage over 200 million acres of national parks, forests and grasslands, many of which harbor infestations of invasive species. It also has the authority to provide technical and financial assistance (primarily for insect, disease, and invasive plant suppression) for all the Nation's 731 million acres of forest lands, including urban, State, private, and tribal lands.

In 1999, Executive Order (EO) 13112 established the National Invasive Species Council (NISC), co-chaired by the Secretaries of the Interior, Agriculture, and Commerce. NISC members include the Secretaries of Transportation, State, Defense, Homeland Security, Treasury, and Health and Human Services; the Administrators of the Environmental Protection Agency and the National Aeronautics and Space Administration; as well as the Director of the U.S. Agency for International Development and the U.S. Trade Representative. NISC was charged with providing coordination, planning and overall leadership for federal invasive species programs and reaching out to state, tribal, local and private partners.

Coordination is not only critical between agencies of the Federal government, but also with state and local entities. Recent efforts to create national management frameworks have helped coordinate US Forest Service (USFS), Animal Plant health Inspection Service (APHIS), National Association of State Foresters (NASF), National Plant Board (NPB), and their missions, expertise and available resources to effectively respond to three priority invasive species threats: Sudden Oak Death, Emerald Ash Borer, and Thousand Canker Disease.

Actions at the local level are also critical. I'd like to highlight some efforts in my home state of West Virginia. The Potomac Highlands Cooperative Weed and Pest Management Area (CWPMA) is a partnership between federal, state, and local agencies, community associations, non-profit organizations, and private land owners aimed at coordinating efforts and programs for addressing the threat of invasive species. The mission of Potomac Highlands CWPMA is the prevention and management of invasive species in the headwaters region of the South Branch of the Potomac River in West Virginia and Virginia. They are dedicated to decreasing the impacts of invasive species on native plant and animal communities, public and private forests, private and agricultural lands, and local economies through public awareness, education, professional improvement and environmental awareness. Projects include volunteer work days, landowner education, and youth events at schools.

Other efforts like those in Georgia are essential in the fight against invasive species such as Cogon grass. The Georgia "Cogon grass Task Force" has provided training to resource professionals throughout the state, and its educational campaign continues to help landowners identify and remove the plant. The Georgia Forestry Commission spearheaded an effort to bring a total of 23 state, federal and private partners to establish the entire state of Georgia as a Cooperative Weed Management Area for Cogon grass in May 2008. The combined effort of this group should have far reaching impacts to help educate the public about Cogon grass as well as help locate all infested sites.

Legislation and program implementation is needed to increase the nation's current protection system for invasive species, which is currently piecemeal and lacks adequate rigor and comprehensiveness, virtually ensuring that invasive species will continue to arrive and spread. Federal government involvement with states is critical as specific federal legislation (e.g. inter-state commerce, plant protection) limit certain state actions. A successful forest invasive species prevention and control program must address the complexity and wide-ranging agency and community needs at the regional, state, and local level. The National Association of State Foresters believes that:

- Federal agencies (e.g., USDA Animal and Plant Health Inspection Service) should coordinate with appropriate state agencies as well as their traditional state agricultural agency partners in addressing forest invasive pests. If desired, state agencies should be provided the option to have a lead role in deciding what programs, regulations and initiatives are needed and best suited to protect forest resources within their respective states. In most cases state agencies have better knowledge and contacts with local stakeholders and community-based organizations that have developed tactics and programs to combat invasive species at the local level (e.g., county representatives, utilities).
- Where federal forests dominate ownership at a local or regional level, Federal Agencies should partner strategically with state forestry agencies to minimize the spread of invasives between adjoining private or state forest land and identify opportunities for cost-effective treatment.
- A federal program should include initiatives that are non-regulatory and incentive driven, support and build capacity at the state, regional, and community level, and encourage voluntary cooperation of affected private entities and communities.
- Investments to support local, state, and regional partnerships, which are prepared to take immediate action against known priority invasive species, will provide valuable lessons for others and promote innovations and efficiencies in protection and public outreach strategies. By sharing their progress, these partnerships will, in turn, help identify the policy and legal obstacles to success as well as build a constituency for more effective invasive species prevention and control programs in other areas.

NASF recommends the following strategies to be considered in any comprehensive invasive species legislation:

- Establishment of a state-level rapid response capacity that can quickly eradicate priority forest invasive species.
- Non-regulatory and incentive-driven national programs, with specific focus on encouraging voluntary cooperation.
- Adoption and enforcement of workable national regulatory programs to address key pathways such as firewood movement where non-

regulatory approaches are not in place, effective or, simply, to serve as a "backstop" for those voluntary approaches.

- Support efforts to enhance forest invasive species response, management and restoration in areas and communities that have been impacted by harmful non-native forest insects and diseases.
- Development of procedures that resolve jurisdictional and other disputes in an effort to improve coordination of Federal agencies, and between Federal and State agencies. A network of partners is needed with agreed upon authorities, responsibilities, and roles.
- Increase research capacity in efforts to quickly assess impacts of potential invasive species; identify, test, and deploy bio-control agents, develop management tools for mitigation and suppression, and genetic and breeding programs designed to enhance resistance of high priority tree species; and
- An increased percentage of funds delegated to the states and their cooperating entities. In most cases, state agencies have better knowledge and contacts with local stakeholders and community-based organizations.

I appreciate the opportunity to appear before the Subcommittee today to offer perspectives shared by state foresters regarding the impacts of invasive species on the nation's forests. I would like to thank the Subcommittee for its continued leadership and support of active, sustainable management of all forest lands.

INDEX

B

C

D

I

J

K

L

M

N

Q

R

S

T

U

V

W

Y